# Fundamental Theories of Physics

## Volume 168



For further volumes:
http://www.springer.com/series/6001

Virginia Velma Fernández • Waldyr A. Rodrigues Jr.

# Gravitation as a Plastic Distortion of the Lorentz Vacuum

Springer

Dr. Virginia Velma Fernández
Universidad de Buenos Aires
Fac. Ingenieria
Depto. Fisica
Av. Paseo Colón 850
1063 Buenos Aires
Argentina
vfernandez@fi.uba.ar

Prof. Dr. Waldyr A. Rodrigues Jr.
Universidade Estadual de
Campinas
Instituto de Matemática
Estatística e
Computação Científica
Campinas São Paulo
Brazil
walrod@ime.unicamp.br

ISBN 978-3-642-13588-0 e-ISBN 978-3-642-13589-7
DOI 10.1007/978-3-642-13589-7
Springer Heidelberg Dordrecht London New York

Library of Congress Control Number: 2010935173



*Cover design*: deblik, Berlin

Printed on acid-free paper

Springer is part of Springer Science+Business Media (www.springer.com)

# Preface

In this book we present a theory of the gravitational field where this field (a kind of 'square root' of $\boldsymbol{g}$) is represented by a $(1,1)$-extensor field $\boldsymbol{h}$ describing a plastic distortion of the Lorentz vacuum (a complex substance that lives in a Minkowski spacetime) due to the presence of matter. The field $\boldsymbol{h}$ distorts the Minkowski metric extensor $\eta$ generating what may be interpreted as an effective Lorentzian metric extensor $\boldsymbol{g} = \boldsymbol{h}^{\dagger}\eta\boldsymbol{h}$. Besides that, $\boldsymbol{h}$ permits the introduction of different kinds of parallelism rules on the world manifold, which may be interpreted as distortions of the parallelism structure of Minkowski spacetime and which may also have non null curvature and/or torsion and/or nonmetricity tensors. Therefore, we have different possible effective geometries which may be associated with the gravitational field and thus its description by a Lorentzian geometry is only a possibility, not an imposition from Nature. Moreover, we present with enough details the theory of multiform functions and multiform functionals which is the main new ingredient permitting us to write a Lagrangian for $\boldsymbol{h}$ successfully and to obtain its equations of motion, that results in our theory being equivalent to Einstein field equations of General Relativity (for all those solutions where the manifold $M$ is diffeomorphic to $\mathbb{R}^4$). However, in our theory, differently from the case of General Relativity a trustworth energy-momentum conservation law and an orbital plus spin angular momentum conservation law exist. We also express the results of our theory in terms of the gravitational potentials $\mathfrak{g}^{\mu} = \boldsymbol{h}^{\dagger}(\vartheta^{\mu})$ where $\{\vartheta^{\mu}\}$ is an orthonormal basis of Minkowski spacetime (representing the ground state of the Lorentz vacuum), in order to have results which may be easily expressed with the theory of differential forms. The nice Hamiltonian formalism for our theory (formulated in terms of the potentials $\mathfrak{g}^{\mu}$) is also discussed with details. The book contains also several important Appendices that complement the material in the main text.

The authors are grateful to Dr. E. A. Notte-Cuello, Dr. A. M. Moya, Dr. E. Capelas de Oliveira, Dr. R. da Rocha, Dr. Q. A. G. Souza and Dr. J. Vaz Jr. for many useful discussions on the subject of this book, and want to thank also Dr. J. Dunning-Davies for his improvement of our English.

# Contents

# 1 Introduction

ABSTRACT In this chapter we first recall the concept of a geometrical space structure that may be defined on a manifold $M$ once we introduce different metrical fields and different connections on it. Then, we brielfly recall the concepts of curvature, torsion and nonmetricity of a general connection[1] on $M$. Next we recall the concepts of flat space and affine spaces. After that, we explain with details and many examples[2] the crucial difference between the concept of curvature of a connection defined on $M$ and the fact that $M$ can eventually be a bent hypersurface in an Euclidean (or pseudo-Euclidean) space with an appropriate number of dimensions. After that, we present the main reason for the proposal of our theory, namely the fact that in Einstein's General Relativity Theory (*GRT*) there are no trustworth energy-momentum and angular momentum conservation laws[3]. At the end of the chapter, we describe the contents of the other chapters and appendices briefly.

## 1.1 Geometrical Space Structures, Curvature, Torsion and Nonmetricity Tensors

In this book we present a theory of the gravitational field where this field is described by a distortion field $\boldsymbol{h}$ which exists and interacts with the matter fields in Minkowski spacetime. It describes a plastic deformation of the Lorentz vacuum understood as a real physical substance[4]. The distortion field is represented in a mathematical formalism —called the multiform and extensor calculus on manifolds (*MECM*)— by a $(1,1)$-extensor field and as we are going to see, $\boldsymbol{h}$ is a kind of square root

[1]Those concepts are presented with details in Chapter 4.
[2]See also Appendices A and B.
[3]Which, of course, is not the case in our theory.
[4]We recall that deformations of a medium may be of the elastic or plastic type [1]. Later we will explain the difference between those two types of deformation.

V.V. Fernández and W.A. Rodrigues Jr., *Gravitation as a Plastic Distortion of the Lorentz Vacuum*, Fundamental Theories of Physics 168, DOI 10.1007/978-3-642-13589-7_1,

of a $(1,1)$-extensor field $\boldsymbol{g}$ which is directly associated to the metric tensor $\mathbf{g}$ which is well known to represent important aspects of the gravitational field in Einstein's *GRT*. Before going into the details, we will digress in order to present the main motivations for our enterprise.

We start by recalling that in *GRT*, the gravitational field has a status which is completely different from the other physical fields. Indeed, in *GRT* the gravitational field is interpreted as aspects of a *geometrical* structure of the spacetime manifold. More precisely, we have that each gravitational field generated by a given matter distribution (represented by a given energy-momentum tensor) is represented by a pentuple $(M, \mathbf{g}, D, \tau_{\mathbf{g}}, \uparrow)$, where[5]:

**(mi)** $M$ is a 4-dimensional Hausdorff manifold which is paracompact, connected and noncompact,

**(mii)** $\mathbf{g} \in \sec T_2^0 M$ is a tensor of signature $-2$, called a *Lorentzian metric*[6] on $M$,

**(miii)** $D$ is the Levi-Civita connection of $\mathbf{g}$,

**(miv)** $M$ is oriented by the volume element $\tau_{\mathbf{g}} \in \sec \bigwedge^4 T^*M$ and the symbol $\uparrow$ means that $M$ is also time oriented.

*GRT* supposes that particles and fields are described by some special structures[7] on the manifold $M$. Particles are described by triples $\{(m, q), S, \sigma\}$ where $m$ $(0 \leq m < \infty)$ is the particle mass, $q$ is a parameter called the electric charge[8], $S$ is the particle's spin[9], and $\sigma$ is a regular curve called the world line of the particle.[10]

The different physical fields are modelled by special sections of the tensor and spinor bundles over the basic manifold $M$.[11]

The world lines that represent the history of particles, as well as the dynamics of the tensor and spinor fields which describe the physical fields, are mathematically described by some system of differential equations (in general called the *equations of motion*) in which the metric tensor $\mathbf{g}$ appears.

The energy-momentum content of a system of particles and fields is described by their respective *energy-momentum tensors* $\mathbf{T}_{\mathrm{p}} \in \sec T_0^2 M$ and $\mathbf{T}_{\mathrm{f}} \in \sec T_0^2 M$. As is well known [2, 6], those objects appear in the second term of Einstein's field equation, whereas in the first term of that equation there appears a tensor called the *Einstein tensor*.

Einstein's equation may be written (after a convenient choice of a local chart on $U \subset M$) as a system of ten nonlinear partial differential equations which contains a certain combination of the first and second order partial derivatives of the components $g_{\mu\nu}$ of the metric tensor.

Einstein's equation is sometimes described pictorially [6] by saying that "the geometry says how the matter must *move* and,in its turn matter says how the geometry must *curve*". This pictorial description induces novices on the subject to imagine that the curvature concept used in *GRT* means that the *manifold* $M$, which models spacetime is some kind of *bent* hypersurface (a 4-brane living in a $(4+p)$-dimensional pseudo-Euclidean space)[12]. It is then necessary, in order to appreciate

---

[5]In fact a gravitational field is defined by an equivalence class of pentuples, where $(M, \mathbf{g}, D, \tau_{\mathbf{g}}, \uparrow)$ and $(M', \mathbf{g}', D', \tau'_{\mathbf{g}}, \uparrow')$ are said to be equivalent if there is a diffeomorphism $\mathtt{h} : M \to M'$, such that $\mathbf{g}' = \mathtt{h}^* \mathbf{g}$, $D' = \mathtt{h}^* D$, $\tau'_{\mathbf{g}} = \mathtt{h}^* \tau_{\mathbf{g}}$, $\uparrow' = \mathtt{h}^* \uparrow$, (where $\mathtt{h}^*$ here denotes the pullback mapping). For more details, see, e.g., [2,3,4].

[6]The field $\mathbf{g}$ in *GRT* obeys Einstein's field equation, to be recalled below.

[7]Particles and fields are to be interpreted in future developments of our theory respectively as different kinds of defects and distortions of the Lorentz vacuum.

[8]It seems to be an experimental fact that if $e$ denotes the charge of the electron, then $q = ne$, $n \in \mathbb{Z}$ for all leptons, mesons and baryons and also that $q$ can take the values $\pm\frac{1}{3}e$ or $\pm\frac{2}{3}e$ for quarks.

[9]For details of how to characterize the particle's spin in the Clifford bundle formalism see e.g., [5].

[10]If $m > 0$ the particle is called a *bradyon* and in this case $\sigma$ is a *timelike* curve pointing into the future. If $m = 0$ the particle is said to be a luxon and in this case $\sigma$ is a *lightlike* curve.

[11]A description of how this can be done using the Clifford bundle formalism may be found in [4].

the developments of this book, to correct such an idea by explaining in a rigorous way and by illustrating with examples what the concept of curvature used in *GRT* really means, and also to realize that the use of this concept in the formulation of certain gravitational theories (including *GRT*) is no more than a coincidence.

We recall that, from a mathematical point of view a general differential manifold $M$ by itself possesses only a topological structure and a differential structure of the charts of its maximal atlas. It must be clear to start, that the topological structure of $M$ is not fixed in *GRT* and indeed the topology of $M$ is introduced 'by hand' in specific problems and situations [2,7]. On the other hand, any given $M$ may support, in general, many distinct *geometrical structures.*

In what follows we say that a pair $(M, \nabla)$, where $M$ is a general differential manifold and $\nabla$ is an arbitrary connection on $M$ is a space endowed with a parallelism rule. We say that the triple $(M, \mathbf{g}, \nabla)$ is a *geometric space structure* (*GSS*) and that $\mathbf{g}$ is a general metric tensor on $M$. The signature of a $\mathbf{g}$ is arbitrary. So, even in the case where $(M, \mathbf{g}, \nabla)$ is part of a Lorentzian spacetime structure and $\mathbf{g}$ has signature $-2$ we can have another *GSS* $(M, \mathbf{g}', \nabla)$ where $\mathbf{g}'$ has another signature[13].

Now, the objects that characterize a *GSS* are the following tensor fields[14]:

**(egi)** *nonmetricity of* $\nabla$, $\mathfrak{A} \in T_3^0 M$,

$$\mathfrak{A} = \nabla \mathbf{g}. \tag{1.1}$$

**(egii)** *torsion of* $\nabla$, represented by a tensor field $\Theta \in T_2^1 M$,
**(egiii)** the *Riemannian curvature* of $\nabla$, represented by a tensor field $\mathfrak{R} \in T_3^1 M$.

A geometric space structure is said to be a:

**(a)** Riemann-Cartan- Weyl *GSS* iff $\mathfrak{A} \neq 0, \Theta \neq 0, \mathfrak{R} \neq 0$,
**(b)** Riemann-Cartan *GSS* iff $\mathfrak{A} = 0, \Theta \neq 0, \mathfrak{R} \neq 0$,
**(c)** Riemann *GSS* iff $\mathfrak{A} = \Theta = 0, \mathfrak{R} \neq 0$.

A general *GSS* $(M, \mathbf{g}, \nabla)$ such that $\nabla \mathbf{g} = 0$ will be called a metric compatible geometric space structure (*MCGSS*)

## 1.2 Flat Spaces, Affine Spaces, Curvature and Bending

A *MCGSS* $(M, \mathbf{g}, \nabla)$ such that $\mathfrak{A} = \Theta = \mathfrak{R} = 0$ is said to be a *globally flat* Riemann *MCGSS*. We denote by $\mathcal{A}^n$ a *MCGSS* where the parallelism rule is such that $\Theta = \mathfrak{R} = 0$ and where $M \simeq \mathbb{R}^n$ is isomorphic to a (real) affine space of the same dimension.

As it is well known [8], a (real) affine space has as a fundamental property an operation called the difference between two given points, such that if $p$, $q \in \mathcal{A}^n$ (or $p, q \in \mathbb{E}^n$) then $(p - q) \in \mathbf{V}$, where $\mathbf{V}$ is a real $n$-dimensional vector space. So, we may say, in a pictorial way, that an affine space is a vector space from which the origin has been stolen. Given an arbitrary point $o \in \mathcal{A}^n$ (said to be the *origin*), the object $(p - o) \in \mathbf{V}$ is appropriately called the position vector of $p$ relative to $o$. This operation permits us to define an *absolute parallelism* rule (denoted $\uparrow\uparrow$) in an affine space. Given two arbitrary points $o$,$o' \in \mathcal{A}^n$, consider the set of all tangent vectors at those two points,

---

[12]Indeed, this is typical of many drawings appearing in textbooks and expository books for laymen.
[13]In particular, in our theory we will see that we shall need to introduce a $\mathbf{g}'$ with signature $+4$.
[14]The definitions of those objects will be recalled in Chapter 3.

i.e., $\{(p-o), p \in \mathcal{A}^n\}$ and $\{(p'-o'), p' \in \mathcal{A}^n\}$. Then we say that a tangent vector at $o$, say $(p-o)$ is parallel to a tangent vector at $o'$, say $(p'-o')$ if being $(p-o) = \mathbf{v} \in \mathbf{V}$ and $(p'-o') = \mathbf{v}' \in \mathbf{V}$ we have $\mathbf{v} = \mathbf{v}'$. This parallelism rule permits us to identify the pair $(\mathcal{A}^n, \upuparrows)$ with a parallelism structure $(M \simeq \mathbb{R}^n, \nabla^{\upuparrows})$ where $\nabla^{\upuparrows}$ is a connection, called the Euclidean connection and defined as follows. Let $\{\mathbf{x}^i\}$, $i = 1, 2, ..., n$ be a set of Cartesian coordinates for $M(\simeq \mathbb{R}^n)$. Consider the global coordinate basis $\{e_i\}$ for $TM$ where $e_i := \partial/\partial \mathbf{x}^i$. Then, put

$$\nabla^{\upuparrows}_{e_j} e_i = 0, \quad \forall i, j = 1, 2, ..., n. \tag{1.2}$$

We define a Euclidean *MCGSS* as a triple $\mathbb{E}^n = (\mathbb{R}^n, \mathbf{g}, \nabla^{\upuparrows})$ where $\mathbf{g}$ defines the standard Euclidean scalar product in each tangent space $T_pM$ by $\mathbf{g}|_p\left(e_i|_p, e_j|_p\right) = \delta_{ij}$. For future reference, we say that if the signature[15] of $\mathbf{g}$ is $(1, n-1)$, i.e. $\mathbf{g}|_p\left(e_i|_p, e_j|_p\right) = \eta_{ij}$, then the matrix $(\eta_{ij})$ with entries $\eta_{ij}$ is the diagonal matrix diag$(1, -1, ..., -1)$.

Now, from the classification of the *GSS* structures given above it is clear that the Riemannian curvature does not refer to any *intrinsic* property of $M$, but refers to a *particular* parallelism structure (i.e., a property of a particular connection $\nabla$) that has been defined on $M$. Keeping this in mind is crucial, because very often we observe confusion between the concept of Riemannian curvature and the concept of *bending*[16] of hypersurfaces embedded in a higher dimensional Euclidean *MCGSS*, something that may lead to equivocated and even psychedelic interpretations concerning the physical and mathematical contents of *GRT*.

So, before proceeding, we should recall some examples, which we hope will clarify the difference between *bending* and *curvature*.

Consider a 2-dimensional cylindrical surface $\mathcal{C}$ embedded in $\mathbb{E}^3$. Using as metric on $\mathcal{C}$ the pullback of the Euclidean metric of $\mathbb{E}^3$ and defining on $\mathcal{C}$ a connection which is the pullback of the connection $\nabla^{\upuparrows}$ of $\mathbb{E}^3$ (which happens to be the Levi-Civita connection of the induced metric), it is possible to verify without difficulties that the Riemannian curvature of $\mathcal{C}$ is *null*, i.e., the cylinder $\mathcal{C}$ is *flat* according to its Riemannian curvature. However, a cylinder $S^1 \times \mathbb{R}$ living in $\mathbb{E}^3$ is very different topologically from any plane $\mathbb{R}^2$ which lives in $\mathbb{E}^3$ and which *may* have zero curvature tensor *if* it is part of a *MCGSS* (call it ${}^p\mathbb{E}^2$), where the metric on it is the pullback of the Euclidean metric of $\mathbb{E}^3$ and its connection is the pullback of the connection $\nabla^{\upuparrows}$ of $\mathbb{E}^3$, (which happens to be the Levi-Civita connection of the induced metric). As we already said (see Footnote 16), it is the *shape tensor* the mathematical object that helps to characterize distinct $r$-dimensional differential manifolds (with identical properties concerning the metric and the connection structures) when they are viewed as hypersurfaces embedded in an Euclidean (or pseudo-Euclidean) GSS of appropriate dimension $n > r$, and of course, the shape tensors of $\mathcal{C}$ and $\mathbb{R}^2$ (as part of the *MCGSS* ${}^p\mathbb{E}^2$) are very different [11].

Another very instructive example showing the crucial distinction between curvature and bending is the following [12]:

**(A)** Consider the torus $\mathcal{T}_1$ as a surface embedded in $\mathbb{E}^3$ and represented, e.g., in the canonical coordinates of $\mathbb{E}^3$ by the equation

$$(x^2 + y^2 + z^2 + 3)^2 = 16(x^2 + y^2). \tag{1.3}$$

---

[15] A metric with signature $(n-1, 1)$ is also called pseudo-euclidean. Note also that mathematical textbooks define the signature of a metric $(p, q)$ as the number $s = p - q$.

[16] We recall that the bending of a $r$-dimensional hypersurface $S$ considered as a subset of points of an euclidean (or pseudo-eucldean) *GSS* of appropriated dimension [9] $n$ is characterized by the so called *shape tensor* [10,11].

Of course, $\mathcal{T}_1$ is a 2-dimensional manifold. If we consider on $\mathcal{T}_1$ a *MCGSS* structure where the metric on $\mathcal{T}_1$ is the pullback of the Euclidean metric of $\mathbb{E}^3$ and its connection is the pullback of the connection $\nabla^{\upuparrows}$ of $\mathbb{E}^3$ (which results to be the Levi-Civita connection of the induced metric) then the Riemann curvature tensor of that pullback connection on $\mathcal{T}_1$ is *non null*, as is easy to verify.

**(B)** Now, the subset $\mathcal{T}_2$ of $\mathbb{E}^4$ given by,

$$\mathcal{T}_2 = \{x \in \mathbb{E}^4 \mid (x^1)^2 + (x^2)^2 = (x^3)^2 + (x^4)^2 = 1\}, \tag{1.4}$$

is diffeomorphic to the torus $\mathcal{T}_1$ defined in **(A)**. Consider the *MCGSS structure defined on* $\mathcal{T}_2$ where the metric is the pullback of the Euclidean metric of $\mathbb{E}^4$ and its connection is the pullback of the connection $\nabla^{\upuparrows}$ of $\mathbb{E}^4$. It can be easily verified that such *MCGSS* is globally flat, i.e., $\Theta = \mathfrak{R} = 0$.

Those two examples make it clear that we should not confuse curvature with bending. Indeed, these examples show that a given manifold when interpreted as an $r$-dimensional hypersurface embedded in an euclidean *MCGSS* of appropriate dimension $n > r$, may, or may not, be Riemann curved (even if that *MCGSS* has as its metric tensor and its connection, respectively the pullback of the metric tensor and of the connection of $\mathbb{E}^n$). In particular, keep in mind that a 2-dimensional torus in order to be part of a flat *MCGSS* (with a Levi-Civita connection of the pullback metric), must be embedded in a euclidean *GSS* with more dimensions than in the case in which it is part of a non flat *MCGSS* (with a Levi-Civita connection of the pullback metric).[17]

We give yet another example -which Cartan showed to Einstein in 1922 when he visited Paris [13,14]- to clarify (we hope) the question of the curvature of a given *MCGSS*. Consider the punctured sphere $\mathring{S}^2 = \{S^2 -$ north and south poles$\} \subset \mathbb{E}^3$. $\mathring{S}^2$ is clearly a bent surface embedded in $\mathbb{E}^3$. As it is well known, it is a surface with non zero Riemann curvature tensor when interpreted as part of a *MCGSS* $\{\mathring{S}^2, \mathbf{g}, D\}$, where the metric $\mathbf{g}$ and the connection $D$ are respectively the pullback of the metric and the connection of the *MCGSS* $\mathbb{E}^3$.

Now, besides the *MCGSS* $\{\mathring{S}^2, \mathbf{g}, D\}$, it is possible to define on $\mathring{S}^2$ a Riemann-Cartan *MCGSS* $\{\mathring{S}^2, \mathbf{g}, \nabla\}$, where $\mathbf{g}$ is as before and $\nabla$ is a metric compatible Riemann-Cartan connection such that its Riemann curvature is null, but its torsion tensor is *non null*. $\nabla$ is called in [15] the navigator connection, and the denomination Columbus connection is also sometimes used. In [4] $\nabla$ has been appropriately called the Nunes connection[18]. The parallelism rule defining $\nabla$ is very simple (see Figures 2 and 3 in Appendix B). Given a vector $\mathbf{v}$ at $p$, let $\alpha$ be the angle it makes with the tangent vector to the latitude lines that pass through $p$. Then, $\mathbf{v}$ is said to be parallel transported according

---

[17]We can also put on a torus living in $\mathbb{E}^3$ a non metric compatible connection with null torsion and curvature torsions or yet a metric compatible connection with non null torsion tensor and null curvature torsion. See Appendix A.

[18]Pedro Salacience Nunes (1502–1578) was one of the leading mathematicians and cosmographers from Portugal during the Age of Discoveries. He is well known for his studies in Cosmography, Spherical Geometry, Astronomic Navigation, and Algebra, and particularly known for his discovery of the loxodromic curves and the nonius. Loxodromic curves, also called rhumb lines, are spirals that converge to the poles. They are lines that maintain a fixed angle with the meridians. In other words, loxodromic curves are directly related to the construction of the Nunes connection. A ship following a fixed compass direction travels along a loxodromic, this being the reason why Nunes connection is also known as navigator connection. Nunes discovered the loxodromic lines and advocated the drawing of maps in which loxodromic spirals would appear as straight lines. This led to the celebrated Mercator projection, constructed along these recommendations. Nunes invented also the Nonius scales, which allow a more precise reading of the height of stars on a quadrant. The device was used and perfected at the time by several people, including Tycho Brahe, Jacob Kurtz, Christopher Clavius and further by Pierre Vernier, who in 1630 constructed a practical device for navigation. For some centuries, this device was called nonius. During the 19th century, many countries, most notably France, started to call it vernier. More details in http://www.mlahanas.de/Stamps/Data/Mathematician/N.htm.

to the Nunes connection from $p$ to $q$ along any curve containing those points if at a point in the curve, the transported vector makes the same angle $\alpha$ with the tangent vector to the latitude line at that point. It is then possible to verify that, indeed, the Riemann curvature of the Nunes connection is null and its torsion is non null. Details are given in Appendix B where a comparison of the transport rules on $\mathring{S}^2$ according to the Levi-Civita and the Nunes connection is also given. From Appendix B, it is clear that the orthonormal basis $\{\mathbf{e_1} = \frac{1}{r}\partial/\partial\theta, \mathbf{e_2} = \frac{1}{r\sin\theta}\partial/\partial\phi\}$ (where $r$, $\theta$,$\phi$ are the usual spherical coordinates) used in the calculations are not defined on the poles. This is due to the well known fact that $S^2$ does not admit two linear independent tangent vector fields in all its points. Any given vector field on $S^2$ necessarily is zero at some point of $S^2$ (see, e.g., [16]).

A $n$-dimensional manifold $M$ which admits $n$ linearly independent vector fields $\{e_i \in \sec TM$, $i = 1, 2, ..., n\}$ at all its points is said to be *parallelizable*.

A Riemann-Cartan *MCGSS* $\{M, \mathbf{g}, \nabla\}$ which is also parallelizable and such that there exists a set of $n$ linearly independent vector fields $\{e_i \in \sec TM$, $i = 1, 2, ..., n\}$ on $M$ such that

$$\nabla_{e_j} e_i = 0, \ \forall i, j = 1, 2, ..., n \tag{1.5}$$

is said to be a *teleparallel MCGSS*. As may be verified without difficulty, any teleparallel *MCGSS* has null Riemannian curvature tensor but non null torsion tensor.

It is also an easy task to invent examples where a manifold $M$ diffeomorphic to $\mathbb{R}^n$ is part of a *MCGSS* such that its connection has non zero curvature and/or torsion. A simple example is the following. Take an arbitrary global $\mathbf{g}$-orthonormal non coordinate basis for $TM$ ($M \simeq \mathbb{R}^n$) given by $\{\mathbf{f}_i\}$, with $\mathbf{f}_i = f_i^j \partial/\partial \mathrm{x}^j$, and introduce on $M \simeq \mathbb{R}^n$ a connection $\nabla$ such that $\nabla_{\mathbf{f}_j}\mathbf{f}_i = 0$, for all $i, j = 1, 2, ..., n$. It is then easy to verify that the *nonmetricity* of that connection, i.e., $\nabla\mathbf{g} = 0$, the Riemann curvature tensor of $\nabla$ is null but the torsion tensor of $\nabla$ is non null. This last example is important regarding the *interpretation* of the gravitational field in theories that can be shown to be mathematically equivalent to *GRT* (in a precise sense to be disclosed in due course), *not* as an element describing a particular geometrical property of a given *MCGSS*, but as a *physical field* in the sense of Faraday (i.e., a field with ontology similar to the electromagnetic field) existing in Minkowski spacetime. But a reader may ask: is there any *serious* reason for trying such an interpretation for the gravitational field? The answer is yes, and it will be briefly discussed now.

## 1.3 Killing Vector Fields, Symmetries and Conservation Laws

We will start this section by given some specialized names for structures that may represent a spacetime in *GRT* and some of its better known generalizations. Let then $M$ be a 4-dimensional differentiable manifold satisfying the properties required in the definition given above for a (Lorentzian) spacetime of *GRT*. Consider the structure given by the pentuple $(M, \mathbf{g}, \nabla, \tau_{\mathbf{g}}, \uparrow)$, where $\mathbf{g}$ is a Lorentzian metric (with signature $-2$) on $M$ and $\nabla$ is an arbitrary connection on $M$.

We say that the pentuple $(M, \mathbf{g}, \nabla, \tau_{\mathbf{g}}, \uparrow)$ is:

(a) Lorentz-Cartan-Weyl spacetime *iff* $\mathfrak{A} \neq 0, \Theta \neq 0, \mathfrak{R} \neq 0$,

(b) Lorentz-Cartan spacetime *iff* $\mathfrak{A} = 0, \Theta \neq 0, \mathfrak{R} \neq 0$,

(c) Lorentzian spacetime *iff* $\mathfrak{A} = \Theta = 0, \mathfrak{R} \neq 0$,

(d) Minkowski spacetime *iff* $M \simeq \mathbb{R}^4$, $\mathfrak{A} = \Theta = \mathfrak{R} = 0$.

In a Minkowski spacetime we will denote the metric tensor by $\boldsymbol{\eta}$. Moreover, we recall that, in such spacetime, there exists an equivalence class of global coordinate systems for $M \simeq \mathbb{R}^4$ such that if $\{\mathrm{x}^\alpha\}$ is one of these systems, then

$$\boldsymbol{\eta} = \eta_{\alpha\beta} d\mathbf{x}^{\alpha} \otimes \mathbf{x}^{\beta}, \tag{1.6}$$

with $\eta_{\alpha\beta} = \boldsymbol{\eta}(\frac{\partial}{\partial \mathbf{x}^{\alpha}}, \frac{\partial}{\partial \mathbf{x}^{\beta}})$, where $\alpha, \beta = 0, 1, 2, 3$ and the matrix $(\eta_{\alpha\beta})$ is:

$$(\eta_{\alpha\beta}) = \text{diag}(1, -1, -1, -1). \tag{1.7}$$

Now, consider a Lorentzian spacetime. Let $\mathbf{T} \in \sec T^r_s M$ be an arbitrary differentiable tensor field. We say that a diffeomorphism $l : U \to U$ ($U \subset M$) generated by a one-parameter group of diffeomorphisms characterized by the vector field $\xi$ is a *symmetry* of $\mathbf{T}$ iff

$$\pounds_{\xi} \mathbf{T} = 0, \tag{1.8}$$

where $\pounds_{\xi}$ is the Lie derivative [19] in the direction of $\xi$.

This means that the pullback field $l^* \mathbf{T}$ satisfies

$$l^* \mathbf{T} = \mathbf{T}. \tag{1.9}$$

For Lorentzian spacetimes (where $\dim M = 4$) the symmetries of the metric tensor $\boldsymbol{g}$ play a very important role. Indeed, it can be shown that the equation. (see, e.g., [18])

$$\pounds_{\xi} \boldsymbol{g} = 0, \tag{1.10}$$

called the Killing equation, can have the maximum number of ten Killing vector fields, and that the maximum number ten only occurs for Lorentzian spacetimes, which have constant *scalar curvature.* There are only three distinct Lorentzian spacetimes with the maximum number of Killing vector fields: the Minkowski spacetime, the de Sitter spacetime and the anti de Sitter spacetime.

As it is well known, Minkowski spacetime is the mathematical structure used as the 'arena' for the classical theories of fields and particles, and also for the so called relativistic quantum field theories [19].

In the classical relativistic theories of fields and particles it can be shown that there exist trustworthy conservation laws for the energy-momentum, angular momentum and conservation of the center of mass for any system of particles and fields[20]. The proof [21] of that statement depends crucially on the existence of the ten Killing vector fields of the Minkowski metric tensor $\boldsymbol{\eta}$.

This fact is very important because, given a general Lorentzian spacetime modelling, a given gravitational field according to *GRT* and which has a non constant scalar curvature, in general, there is not a sufficient number of Killing vector fields to formulate trustworth conservation laws for the energy-momentum and angular momentum of a given system of particles and fields. Many attempts have been made in order to establish energy-momentum and angular momentum conservation laws in *GRT*. All attempts are based on the fact that although the field $\mathbf{g}$ is part of the spacetime structure, the 'geometry' must also have energy-momentum and angular momentum, and thus there must exist some mathematical objects that assume the roles of its 'energy-momentum tensor'

---

[19] For details, see, e.g., [4,17].

[20] In the relativistic quantum field theories, where fields exist in Minkowski spacetime, there are trustworth conservation laws for any system of interacting fields.

[21] See, e.g., [18].

and its angular momentum tensor, in order to warrant the conservation of the energy-momentum and angular momentum of the system composed of the gravitational and matter fields. However all attempts were (until now) deceptive (unless some additional hypothesis are postulated for the Lorentzian spacetime structure, e.g., that such structure is parallelizable) because all that had been achieved was the association of a series of *pseudo-energy momentum tensors* with the gravitational field [22] and, without additional commentaries from our part, we quote here what the authors of [2] said on this issue:

"*It is a shame to lose the special relativistic total energy conservation in General Relativity. Many of the attempts to resurrect it are quite interesting, many are simply garbage*".

It can be shown that the non-existence of trustworthy energy-momentum and angular momentum for the system of the gravitational field and the matter fields lead to serious inconsistencies. This has already been noted by several scientists, in particular by Levi-Civita (already in 1919!) [23] and discussed by Logunov and collaborators [18] in the eighties of the last century[23]. A modern discussion may be found in [4, 22,25] and the continuous tentatives for finding a solution for the problem within *GRT* (including today's popular approach called quasi-local energy) may be found in [26]. We will discuss briefly such an approach in Section 7, where we study the Hamiltonian formulation of *our* theory.

After more than 85 years of unsuccessful attempts[24] many physicists think that *GRT* needs a revision where the main emphasis must be given to the fact that the gravitational field is a physical field in the *sense* of Faraday, living and interacting with the other physical fields in Minkowski spacetime. However, the formulation of such a theory using as unique ingredients the geometric objects of Minkowski spacetime and the correct representation of the gravitational field (the distortion field $\boldsymbol{h}$) living in that spacetime and in such a way that the resulting theory becomes equivalent to *GRT* in some well defined sense resulted in being a nontrivial task.

In [35] (using the Clifford bundle formalism [4]) a theory of that kind, i.e., one where the gravitational field is described by a physical field existing in Minkowski spacetime and has its dynamics described by a well defined Lagrangian density such that its equations of motion are equivalent to Einstein's field equations (at least for models of *GRT* where the manifold $M$ can be taken as $\mathbb{R}^4$) was proposed. In that theory, it was suggested that the gravitational field $\boldsymbol{g}$ was to be interpreted as a deformation of the Minkowski metric $\boldsymbol{\eta}$ produced by a special $(1,1)$-extensor field $\boldsymbol{h} : \sec \bigwedge^1 T^*M \to \sec \bigwedge^1 T^*M$, said to be the distortion (or deformation) *tensor field.* However, in that instance, those authors were not able to produce a mathematical formalism in which it was possible to work directly with $\boldsymbol{h}$, i.e., writing a Lagrangian density for that field and deducing its equation of motion.

The necessary mathematical formalism to do that job now exists. It is the *MECM* mentioned above. The *MECM* has its origin, to the best of our knowledge, based on some mathematical ideas first proposed by Hestenes and Sobczyc in their book "*Clifford Algebra to Geometrical Calculus*"

[22]The first one to introduce an energy-momentum pseudo-tensor for the gravitational field was Einstein [20]. However, it is possible to introduce an infinity of distinct energy-momentum pseudo-tensors for the gravitational field in *GRT*. The mathematical reasons for this manifold possibility may be found, e.g., in [4,21,22].

[23]A simple way to understand the origin of the issue has already been clearly stated by Schröndinger, who in [24] observed that in a general Lorentzian spacetime you can not even define the momentum of a pair of particles when they are at events say, $\mathfrak{e}_1$ and $\mathfrak{e}_2$ because vectors at the tangent spaces $T_{\mathfrak{e}_1}M$ an $T_{\mathfrak{e}_2}M$ cannot be summed.

[24]Besides Logunov and collaborators we quote here also Feynmann [27] , Schwinger [28], Weinberg [29], Rosen [30] and [31,32,33,34].

[10]. Those ideas have been investigated in [36,37,38], attaining maturity in [39,40,41,42,43,44,45,46, 47,48,49]. In particular, in [45] the theory of derivatives of functionals of extensor fields necessary for the development of the present discussion has been introduced in a rigorous way. That crucial concept will be recalled (with several examples) in Chapter 3. Utilizing *MECM*, presented with enough details in Chapter 4, we will show that it is, indeed, possible to present a theory of the gravitational field as the theory of a field $\boldsymbol{h}$ existing on Minkowski spacetime, and describing a *plastic deformation* of the Lorentz vacuum. The Lorentz vacuum is in our view a real and very complex physical substance[25] which exists in an arena mathematically described by Minkowski spacetime[26]. When the Lorentz vacuum is in its ground state, it is described by a trivial distortion field $\boldsymbol{h} = \mathrm{id}$, but when it is disturbed due to the presence of what we call matter fields, it is described by a nontrivial distortion field $\boldsymbol{h}$ of the *plastic type*, which as we shall see in Section 5, satisfies a well defined field equation, which are (Chapter 6) equivalent to the Einstein field equation for $\boldsymbol{g}$ (at least for models of *GRT* where the manifold $M$ can be taken as $\mathbb{R}^4$). In Chapter 6, we introduce also genuine energy-momentum and angular momentum conservation laws for the system composed of the matter and gravitational fields.

One of the features of the *general* mathematical formalism used in this book, and developed in Chapters 2 to 4, is a formulation of a theory of covariant derivatives on manifolds, where some novel concepts appear, for example, the rotation extensor field $\Omega$, defined in a 4-dimensional *vector manifold* $U_o \subset U$, where $U_o$ represents the points of a given $\mathcal{U} \subset M$ and $U$ is called the *canonical space* (Chapter 4).

Once $U$ is constructed we introduce a Euclidean Clifford algebra denoted by $\mathcal{C}\ell(U,\cdot)$ on $U_o \subset U$ whose *main purpose* is to define an algorithm to perform very sophisticated and indeed nontrivial calculations. Next we introduce on $U_0$ a constant Minkowski metric extensor field $\eta$ and the distortion field $\boldsymbol{h}$, such that the extensor field $\boldsymbol{g} = \boldsymbol{h}^{\dagger}\eta\boldsymbol{h}$ represents in $U_o$ the metric tensor $\boldsymbol{g}$ in $\mathcal{U} \subset M$. Several different *MCGSS* are introduced having $U_o$ as the first member of the triple, and it is shown that some of those structures may be clearly interpreted as a distortion of another one through the action of the distortion extensor $\boldsymbol{h}$. In that way, it will be shown that when $M \simeq \mathbb{R}^4$, in which case we may identify $\mathbb{R}^4 \simeq U_0 \simeq U$, the rotation extensor field that appears in the representation on $U$ of the Levi-Civita connection of $\boldsymbol{g}$ can be written as a functional of $\boldsymbol{h}^{\clubsuit} = \boldsymbol{h}^{-1\dagger}$ and its *vector derivatives* $\cdot\partial\boldsymbol{h}^{\clubsuit}$ and $\cdot\partial \cdot \partial\boldsymbol{h}^{\clubsuit}$. This permits us to write the usual Einstein-Hilbert Lagragian as a functional of $(\boldsymbol{h}^{\clubsuit}, \cdot\partial\boldsymbol{h}^{\clubsuit}, \cdot\partial \cdot \partial\boldsymbol{h}^{\clubsuit})$ and to find directly the equation of motion for $\boldsymbol{h}$, a result that for the best of our knowledge has not been obtained before in a consistent way. It must be said, though, that the rudiments of the multiform and extensor calculus to formulated a theory of the gravitational field as a gauge theory (but where the word gauge does not have the same meaning as the one it has in those field theories which use gauge fields defined as sections of some principal bundles) had been developed in [54,55]. However, despite some very good ideas presented there, those works contain in our opinion some equivocated mathematical concepts which lead to inconsistencies. The gravitational theory in [54,55] has been recently reviewed in [56], but it contains the same equivocated mathematical concepts. So, to distinguish clearly our theory from the one in [54,55], we briefly discuss the main ingredients of that theory in Appendix F and point out explicitly what those inconsistencies are.

[25] For some interesting ideas on the nature of such a substance see [50,51,52,53].

[26] Eventually future experimental developments will show that this hypothesis must be modified, e.g., the substance describing the vacuum may exist in a spacetime with more dimensions that Minkowski spacetime.

For completeness, we also present in Appendix C the derivation of the equations of motion for a theory where the gravitational field is described by two independent fields, a distortion field $\boldsymbol{h}$ and a rotation field $\Omega$. This is the line of ideas (first present in [57]) where it is claimed that $\Omega$ is related to the source of spin[27]. Appendix D contains a proof of the equivalence of two apparently very different expressions for the Einstein-Hilbert Lagrangian density. Appendix E recalls the derivation of the field equations for the potential fields $\mathfrak{g}^{\alpha} = \boldsymbol{h}^{\dagger}(\vartheta^{\alpha})$. Also, Appendix G presents (in order to illustrate once more the philosophy adopted in this book) a model for the gravitational field of a point mass where that field is represented by the nonmetricity of a particular connection. Finally, in Chapter 8, we present our conclusions.

## References

[1] Zorawski, M., *Theorie Mathematiques des Dislocations*, Dunod, Paris, 1967.
[2] Sachs, R. K., and Wu, H., *General Relativity for Mathematicians*, Springer-Verlag, New York 1977.
[3] Rodrigues, W. A., Jr., and Rosa, M. A. F., The Meaning of Time in Relativity and Einstein's Later View of the Twin Paradox, *Found. Phys.* **19**, 705–724 (1989).
[4] Rodrigues, W. A. Jr., and Capelas de Oliveira, E., *The Many Faces of Maxwell, Dirac and Einstein Equations. A Clifford Bundle Approach.* Lecture Notes in Physics **722**, Springer, Heidelberg, 2007.
[5] Rodrigues, W. A. Jr., Vaz Jr., J., and Pavsic, M., The Clifford Bundle and the Dynamics of the Superparticle, *Banach Center Publications, Polish Acad. Sci.* **38**, 295–314 (1996).
[6] Misner, C. M., Thorne, K. S., and Wheeler, J. A., *Gravitation*, W. H. Freeman and Co., San Francesco, 1973.
[7] Eddington, A. S., *The Mathematical Theory of Relativity* (3rd edn.), Chelsea, New York, 1975.
[8] Trautman, A., Foundations and Current Problems of General Relativity, in Trautman, A., Pirani, F. A. E., and Bondi, H., *Lectures on General Relativity*, Brandeis Summer Institute in Theoretical Physics, vol. 1, pp. 1-248 Prentice Hall, Inc., London, 1965.
[9] Clarke, C. J. S., On the Global Isometric Embedding of Pseudo Riemannian Manifolds, *Proc. Roy. Soc. A* **314**, 417-428 (1970).
[10] Hestenes D., and Sobczyk, G., *Clifford Algebras to Geometric Calculus*, D. Reidel Publ. Co., Dordrecht. 1984.
[11] O'Neill, B., *Elementary Differential Geometry*, Academic Press, New York, 1966.
[12] Dodson, C. T. J., *Tensor Geometry*, Springer-Verlag, Berlin, 1991.
[13] Debewer, R., *Élie Cartan- Albert Einstein: Letters on Absolute Parallelism*, Princeton University Press, Princeton, 1979.
[14] Goenner, H. F. M, On the History of Unified Field Theories, *Living Rev. in Relativity* **7**, lrr-2004-2(2004)[http://relativity.livingreviews.org/Articles/lrr-2004-2]
[15] Nakahara, M., Geometry, *Topology and Physics*, Institute of Physics Publ., Bristol, 1990.
[16] Eisenberg M.,and Guy, R., A Proof of the Hairy Ball Theorem, *The Am. Math. Monthly* **86**, 571-574 (1979).

[27] However, it must be said here that, as is clear from the the application of the Lagrangian formalism for multiform fields (including spinor fields) using the multiform and extensor calculus, as developed, e.g., in [4], the source of spin is to be found in the antisymmetric part of the canonical energy-momentum tensor.

[17] Choquet-Bruhat, Y., DeWitt-Morette, C., and Dillard-Bleick, M., *Analysis, Manifolds and Physics* (revised edition), North Holland Publ. Co., Amsterdam, 1982.
[18] Logunov, A. A., and Mestvirishvili, M., *The Relativistic Theory of Gravitation*, Mir Publ., Moscow, 1989.
[19] Bogolubov, N. N., Logunov, A. A., and Todorov, I. T., *Introduction to Axiomatic Quantum Field Theory*, W. A. Benjamin, Inc., Reading, MA, 1975.
[20] Einstein, A., Hamiltonsches Princip und Allgemeine Relativitästheorie, Sizungsberichte der Preussischen Akad. d. Wissenschaften (1916). English translation in Einstein, A. Lorentz, H. A., Weyl, H. and Minkowski, H., The Principle of Relativity, pp.167-173, Dover Publ. Inc., New York, 1952.
[21] Thirring, W. and Wallner, R., The Use of Exterior Forms in Einstein's Gravitational Theory, *Brazilian J. Phys.* **8**, 686-723 (1978).
[22] Notte-Cuello, E. A., and Rodrigues, W. A. Jr., Freud's Identity of Differential Geometry, the Einstein-Hilbert Equations and the Vexatious Problem of the Energy-Momentum Conservation in GR, *Adv. Appl. Cliford Algebras* **19,** 113-145 (2009). [arXiv:0801.2559]
[23] Levi-Civita, T., Mechanics-On the Analytical Expression that Must be Given to the Gravitational Tensor in Einstein's Theory, *Rend. della Reale Accad. dei Lincei*, **26** 381-389, (1917). [www.sjcrothers.plasmaresources.com/Levi-Civita.pdf]
[24] Schrödinger, E., *Space-Time Structure*, Cambridge University Press, Cambridge, 1954.
[25] Benn, I. M., Conservation Laws in Arbitrary Space-times, *Ann. Inst. H. Poincaré*, **XXXVII**, 67-91 (1982).
[26] Szabados, L. B., Quasi-Local Energy-Momentum and Angular Momentum in GR: A Review Article, *Living Reviews in Relativity*, [http://www.livingreviews.org/lrr-2004-4]
[27] Feynman, R. P., Morinigo, F. B. and Wagner, W. G., (edited by Hatfield, B.), *Feynman Lectures on Gravitation*, Addison-Wesley Publ. Co., Reading, MA, 1995.
[28] Schwinger, J., Particles, *Sources and Fields, vol. 1*, Addison-Wesley Publ. Co., Reading, MA, 1970.
[29] Weinberg, S., *Gravitation and Cosmology*, J. Wiley and Sons, Inc., New York, 1972.
[30] Rosen, N., A Bimetric Theory of Gravitation, *Gen. Rel. Grav.* **4**, 435-447(1973).
[31] Gupta S. N, Quantization of Einstein's Gravitational Field: General Treatment, *Proc. Phys. Soc. A* **65**, 608-619 (1952).
[32] Thirring, W., An Alternative Approach to the Theory of Gravitation, *Ann. Phys.* **16**, 96-117 (1961).
[33] Grishchuk, L. P., Petrov, A. N., and Popova, A. D., Exact Theory of the (Einstein) Gravitational Field in an Arbitrary Background Spacetime, *Comm. Math. Phys.* **94**, 379-396 (1984).
[34] Grishchuk, L. P, Some Uncomfortable Thoughts on the Nature of Gravity, Cosmology, and the Early Universe (to appear in *Space Sciences Reviews*), [arXiv:0903.4395]
[35] Rodrigues, W. A., Jr., and Souza, Q. A.G., The Clifford Bundle and the Nature of the Gravitational Field, *Found. of Phys.* **23**, 1465–1490 (1993).
[36] Fernández, V. V., *Distortion and Rotation Extensors in Einstein's Gravitational Theory*, Ph.D. thesis (in Portuguese), IFGW-UNICAMP, 2000.
[37] Moya, A.M., *Lagrangian Formalism for Multivector Fields on Spacetime*, Ph D. thesis (in Portuguese), IMECC-UNICAMP, 1999.
[38] Souza, Q. A. G.,*The Dirac Operator, Riemann-Cartan-Weyl Spaces and the Nature of the Gravitational Field*, Ph.D. thesis, in Portuguese, IFGW-UNICAMP (1992).

[39] Fernández, V. V., Moya, A. M., and Rodrigues, W. A. Jr., Euclidean Clifford Algebra, *Adv. in Appl. Clifford Algebras* **11**(S3), 1-21 (2001). [http://www.clifford-algebras.org/]
[40] Fernández, V. V., Moya, A. M. ,and Rodrigues, W. A. Jr., Extensors, *Adv. Appl. Clifford Algebras* **11**(S3), 23-40 (2001). [http://www.clifford-algebras.org/]
[41] Fernández, V. V., Moya, A. M., and Rodrigues, W. A. Jr., Metric Tensor Vs. Metric Extensors, *Adv. Appl. Clifford Algebras* **11**(S3), 41-48 (2001). [http://www.clifford-algebras.org/].
[42] Moya, A. M, Fernández, V. V., and Rodrigues, W. A. Jr., Metric Clifford Algebra, *Adv. Appl. Clifford Algebras* **11**(S3) 49-68, (2001). [http://www.clifford-algebras.org/]
[43] Moya, A. M, Fernández, V. V., and Rodrigues, W. A. Jr., Multivector Functions of Real Variable, *Adv. Appl. Clifford Algebras* **11**(S3), 69-77, (2001). [http://www.clifford-algebras.org/]
[44] Moya, A. M, Fernández, V. V., and Rodrigues, W. A. Jr., Multivector Functions of Mutivector Variable, *Adv. Appl. Clifford Algebras* **11**(S3), 79-91 (2001). [http://www.clifford-algebras.org/]
[45] Moya, A. M,.Fernández, V. V., and Rodrigues, W. A. Jr., Multivector Functionals, *Adv. Appl. Clifford Algebras* **11**(S3), 93-103 (2001). [http://www.clifford-algebras.org/]
[46] Fernández, V. V., Moya, A. M., and Rodrigues, W. A. Jr., Geometric Algebras and Extensors, *Int. J. Geom. Meth. Math. Phys* **4** , 927-964 (2007). [http://arxiv.org/abs/math.DG/0703090]
[47] Fernández, V. V., Moya, A. M., and Rodrigues, W. A. Jr., Multivector and Extensor Fields in Smooth Manifolds, *Int. J. Geom. Meth. Math. Phys.* **4** , 965-985 (2007). [http://arxiv.org/abs/math.DG/0501559]
[48] Fernández, V. V., Moya, A. M., and Rodrigues, W. A. Jr., Geometric and and Extensor Algebras in the Study of the Differential Geometry of Arbitrary Manifolds, *Int. J. Geom. Meth. Math. Phys.* **4**, 1117-1158 (2007). [http://arxiv.org/abs/math.DG/0703094]
[49] Fernández, V. V., Moya, A. M, da Rocha, R., and Rodrigues, W. A. Jr., Clifford and Extensor Calculus and the Riemann and Ricci Extensor Fields of Deformed Structures (M,$\nabla$', $\eta$) and (M,$\nabla$,g), *Int. J. Geom. Meth. Math. Phys.* **4**, 1159-1172 (2007). [http://arxiv.org/abs/math.DG/0502003]
[50] Laughlin, R.B., A Different Universe: Reinventing Physics from the Bottom, Basic Books, New York, 2005.
[51] Unzicker, A., What Can Physics Learn from Continuum Mechanics, [arXiv:gr-qc/0011064v1]
[52] Unzicker, A., *Teleparallel Space-Time with Defects yields Geometrization of Electrodynamics with Quantized Charges.* [arXiv:gr-qc/9612061v2]
[53] Volovik, G. E., *The Universe in a Helium Droplet*, Clarendon Press, Oxford, 2003.
[54] Lasenby, A., Doran, C., and Gull, S., Gravity I -Gravity V, pp. 171-236, in Baylis, W. E. (ed.), *Clifford (Geometric) Algebras,* Birkhaüser, Boston, 1996.
[55] Doran, C., and Lasenby, A., *Geometric Algebra for Physicists*, Cambridge Univ. Press, Cambridge 2003.
[56] Hestenes, D., Gauge Theory Gravity with Geometrical Calculus, *Found. Phys.* **35**, 903-969 (2005).
[57] Helh, F. W., and Datta, B. K., Nonlinear Spinor Equation and Asymmetric Conection in General Relativity, *J. Math. Phys.* **12**, 798-808 (1967).

# 2

# Multiforms, Extensors, Canonical and Metric Clifford Algebras

ABSTRACT In this chapter, we introduce the concepts of multiforms, extensors, the canonical and metric Clifford algebras. Given a real vector space $V$ with $\dim V = n$, multiforms are elements of a space denoted $\bigwedge V$, each multiform being a sum of non homogeneous $k$-forms $(0 \leq k \leq n)$. We define the exterior product of those objects. Next we introduce a Euclidean product (called canonical product) in $V$ and extend it to the space $\bigwedge V$, defining also the right and left contracted products of multiforms. Those concepts permit us to introduce the canonical Clifford product in $\bigwedge V$. We, thus, get a very powerful algebraic structure, the canonical Clifford algebra of $\bigwedge V$, denoted $\mathcal{C}\ell(V, \cdot)$ that is utilized as a basic calculational tool in the rest of the text. Next, we introduce the concept of extensors, linear mappings from Cartesian products $\bigwedge V \times .... \times \bigwedge V \to \bigwedge V$ and study with details the properties of so-called (1,1)-extensors and some important extensors related to it (in particular its extensions), which will appear in different occasions in the following chapters. In particular, they will play a crucial role in our formulation of the differential geometry of manifolds (Chapter 4) and our theory of the gravitational field (Chapter 5). Equipped with the concept of extensor, we introduce in $\bigwedge V$ a metrical extensor $\boldsymbol{g}$ of arbitrary signature (an object directly associated to a metric tensor $\boldsymbol{g}$ of the same signature) and extend it to $\bigwedge V$. Given a $\boldsymbol{g}$, we can define a corresponding metric Clifford algebra. We introduce a pseudo-Euclidean metric extensor $\eta$ in $V$ and find a (1,1)-extensor $\boldsymbol{h}$ which relates $\eta$ to $\boldsymbol{g}$ (the deformation of $\eta$ by $\boldsymbol{h}$). Moreover, we derive an important formula which permits obtaining $\boldsymbol{h}$ (modulus a Lorentz transformation) in terms of a given $\boldsymbol{g}$. Several remarkable results are then presented, such as the golden rule relating the scalar, contracted and Clifford products associated to different metric extensors. With the golden rule, it becomes possible also to relate easily thorough a nice formula, the Hodge star operators associated with two different metric extensors. The chapter ends with a section presenting a set of useful identities that are used several times in the remainder of the book.

V.V. Fernández and W.A. Rodrigues Jr., *Gravitation as a Plastic Distortion of the Lorentz Vacuum*,
Fundamental Theories of Physics 168, DOI 10.1007/978-3-642-13589-7_2,

## 2.1 Multiforms

Let $\mathbf{V}$ be a vector space ($\dim \mathbf{V} = n$) over the real field $\mathbb{R}$. The dual space of $\mathbf{V}$ is usually denoted by $\mathbf{V}^*$, but in what follows it will be denoted by $V$. The space of $k$-forms ($0 \leq k \leq n$) will be denoted by $\bigwedge^k V$ and we have the usual identifications: $\bigwedge^0 V = \mathbb{R}$ and $\bigwedge^1 V = V$.

A *formal* sum $X = X_0 + X_1 + \cdots + X_k + \cdots + X_n$, where $X_k \in \bigwedge^k V$ will be called a *nonhomogeneous multiform.* The set of all nonhomogeneous multiforms has a natural vector space structure[1] over $\mathbb{R}$ and will be denoted by $\bigwedge V$.

Let $\{\mathbf{e}_k\}$ and $\{\varepsilon^k\}$ be respectively bases of $\mathbf{V}$ and $V$ where the second is the *dual* of the first, i.e., $\varepsilon^k(\mathbf{e}_j) = \delta_j^k$. Then, $\{1, \varepsilon^{j_1}, \ldots, \varepsilon^{j_1} \wedge \ldots \wedge \varepsilon^{j_k}, \ldots, \varepsilon^{j_1} \wedge \ldots \wedge \varepsilon^{j_n}\}$ is a basis for $\bigwedge V$, and since $\dim \mathbf{V} = n$, we have that $\dim \bigwedge V = \binom{n}{0} + \binom{n}{1} + \cdots + \binom{n}{k} + \cdots + \binom{n}{n} = 2^n$.

### 2.1.1 The $k$-Part Operator and Involutions

Let $0 \leq k \leq n$, the linear operator $\langle \rangle_k$ defined by $X \mapsto \langle X \rangle_k := X_k$ ($X_k \in \bigwedge^k V$) will be called the $k$-part operator and $\langle X \rangle_k$ is read as the *$k$-part* of $X$.

The $k$-forms are called *homogeneous multiforms of grade $k$* or sometimes *$k$-homogeneous multiforms.* It is obvious that, for any $k$-homogeneous multiform $X$, we have: $\langle X \rangle_l = X$ if $l = k$, or $\langle X \rangle_l = 0$ if $l \neq k$. Also, any multiform may be expressed as a sum of its $k$-parts, i.e., $X = \sum_{k=0}^{n} \langle X \rangle_k$.

The 0-forms $\langle X \rangle_0$ (i.e., the real numbers), the 1-forms $\langle X \rangle_1$, the 2-forms $\langle X \rangle_2$, etc...., are called scalars, forms, biforms, etc. .., and the $n$-forms $\langle X \rangle_n$ and the $(n-1)$-forms $\langle X \rangle_{n-1}$ are sometimes called pseudo-scalars and pseudo 1-forms [2].

We now introduce two fundamental involutions on $\bigwedge V$.

The linear operator $\hat{}$, defined by $X \mapsto \hat{X}$, such that

$$\left\langle \hat{X} \right\rangle_k = (-1)^k \langle X \rangle_k, \tag{2.1}$$

is called the *conjugation operator* and $\hat{X}$ *is read as the conjugate of* $X$.

The linear operator$\tilde{}$, defined by $X \mapsto \widetilde{X}$, such that

$$\left\langle \widetilde{X} \right\rangle_k = (-1)^{\frac{1}{2}k(k-1)} \langle X \rangle_k,$$

is called the reversion operator and $\widetilde{X}$ is read as the reverse of $X$.

Both operators are involutions, i.e., $\hat{\hat{X}} = X$ and $\widetilde{\widetilde{X}} = X$.

[1] The sum of multiforms and the multiplication of multiforms by scalars are defined by: **(i)** if $X = X_0 + X_1 + \cdots + X_n$ and $Y = Y_0 + Y_1 + \cdots + Y_n$ then $X + Y = (X_0 + Y_0) + (X_1 + Y_1) + \cdots + (X_n + Y_n)$, **(ii)** if $\alpha \in \mathbb{R}$ and $X = X_0 + X_1 + \cdots + X_n$, then, $\alpha X = \alpha X_0 + \alpha X_1 + \cdots + \alpha X_n$.

[2] Such a nomenclature must be used with care, since it may lead to serious confusions if mixed up with the concept of de Rham's pair and impair forms, those latter objects also called by some authors pseudo-forms or twisted forms. See [1] for a discussion on the issue.

### 2.1.2 Exterior Product

The exterior product (or Grassmann product) $\wedge : \bigwedge V \times \bigwedge V \to \bigwedge V$ is defined for arbitrary $X, Y \in \bigwedge V$ as the element $X \wedge Y \in \bigwedge V$ such that

$$\langle X \wedge Y \rangle_k = \sum_{j=1}^{k} \langle X \rangle_j \wedge \langle Y \rangle_{k-j} . \tag{2.2}$$

On the right side of Eq.(2.2), it appears a sum of the exterior product of [3] $j$-form by a $(k-j)$-form. The exterior product (like the sum) is an internal law in the space $\bigwedge V$. It satisfies distributive laws on the left and on the right, the associative law $X \wedge (Y \wedge Z) = (X \wedge Y) \wedge Z$ for $X, Y, Z \in \bigwedge V$ and also, for any $\alpha \in \mathbb{R}$ it holds, $\alpha(X \wedge Y) = (\alpha X \wedge Y) = (X \wedge \alpha Y)$. The first are consequence of the distributive laws of the exterior product of $k$-forms, and the second may be shown to be true, without difficulty. The linear space $\bigwedge V$, equipped with the exterior product is an associative algebra called the *exterior algebra of multiforms.*

We will present now the main properties of the exterior product of multiforms.

**(i)** For all $\alpha, \beta \in \mathbb{R}$ and $X \in \bigwedge V$

$$\alpha \wedge \beta = \alpha\beta, \tag{2.3a}$$
$$\alpha \wedge X = X \wedge \alpha = \alpha X, \tag{2.3b}$$

**(ii)** For all $X_j \in \bigwedge^j V$ and $Y_k \in \bigwedge^k V$

$$X_j \wedge Y_k = (-1)^{jk} Y_k \wedge X_j. \tag{2.4}$$

**(iii)** For all $a \in \bigwedge^1 V$ and $X \in \bigwedge V$

$$a \wedge X = \hat{X} \wedge a. \tag{2.5}$$

**(iv)** For all $X, Y \in \bigwedge V$

$$(X \wedge Y)\hat{} = \hat{X} \wedge \hat{Y}, \tag{2.6a}$$
$$\widetilde{X \wedge Y} = \widetilde{Y} \wedge \widetilde{X}. \tag{2.6b}$$

### 2.1.3 The Canonical Scalar Product

Let us fix on $\mathbf{V}$ an arbitrary basis $\{\mathbf{b}_j\}$. A scalar product of multiforms $X, Y \in \bigwedge V$ may be defined by

$$X \cdot Y := \langle X \rangle_0 \langle Y \rangle_0 + \frac{1}{k!} \langle X \rangle_k (\mathbf{b}^{j_1}, \ldots, \mathbf{b}^{j_k}) \langle Y \rangle_k (b_{j_1}, \ldots, b_{j_k}), \tag{2.7}$$

---

[3] The exterior product of $k$-forms is defined as follows: **(i)** $\alpha, \beta \in \mathbb{R} : \alpha \wedge \beta = \alpha\beta$; **(ii)** $\alpha \in \mathbb{R}$ and $x \in \bigwedge^k V$ with $k \geq 1 : \alpha \wedge x = x \wedge \alpha = \alpha x$, **(iii)** if $X_j \in \bigwedge^j V$ and $X_j \in \bigwedge^k V$ with $j, k \geq 1$ then $X_j \wedge Y_k = \frac{(j+k)!}{j!k!} \mathcal{A}(x \otimes y)$ where $\mathcal{A}$ is the well known antisymmetrization operator of ordinary linear algebra. Keep in mind when reading texts on the subject that eventually another definition of the exterior product (not equivalent to the one used here) may be being used, as e.g., in [2].

where, in order to utilize Einstein's sum convention rule, we wrote $b_{j_i} := \mathbf{b}^{j_i}$, $i = 1, 2, ..., k$.

The product defined by Eq.(2.7) (obviously associated with the basis $\{\mathbf{b}_j\}$) is a well defined scalar product on the vector space $\bigwedge V$. It is symmetric ($X \cdot Y = Y \cdot X$), it is linear, i.e., ($X \cdot (Y + Z) = X \cdot Y + X \cdot Z$), it satisfies ($(\alpha X) \cdot Y = X \cdot (\alpha Y) = \alpha(X \cdot Y)$) and it is nondegenerated (i.e., $X \cdot Y = 0$ for all $X$, implies $Y = 0$). Even more, that product is positive definite, i.e., $X \cdot X \geq 0$ for all $X$, and $X \cdot X = 0$ iff $X = 0$.

The scalar product of multiforms, which we just introduced, will be called the *canonical scalar product*, despite the fact that for its definition it is necessary to choose an arbitrary basis $\{\mathbf{b}_j\}$ of $\mathbf{V}$.

Some important properties of the canonical scalar product are:

**(i)** For all scalars $\alpha, \beta \in \mathbb{R}$

$$\alpha \cdot \beta = \alpha\beta, \tag{2.8}$$

**(ii)** For all $X_j \in \bigwedge^j V$ and $Y_k \in \bigwedge^k V$

$$X_j \cdot Y_k = 0, \text{ if } j \neq k. \tag{2.9}$$

**(iii)** For any two simple $k$-forms, say $v_1 \wedge \ldots \wedge v_k$ and $w_1 \wedge \ldots \wedge w_k$, we have:

$$(v_1 \wedge \ldots \wedge v_k) \cdot (w_1 \wedge \ldots \wedge w_k) = \begin{vmatrix} v_1 \cdot w_1 & \cdots & v_1 \cdot w_k \\ \vdots & & \vdots \\ v_k \cdot w_1 & \cdots & v_k \cdot w_k \end{vmatrix}, \tag{2.10}$$

where $|\ |$ denotes here, the classical determinant.

**(iv)** For all $X, Y \in \bigwedge V$

$$\hat{X} \cdot Y = X \cdot \hat{Y}, \tag{2.11a}$$
$$\widetilde{X} \cdot Y = X \cdot \widetilde{Y}. \tag{2.11b}$$

**(v)** Let $(\{\varepsilon^j\}, \{\varepsilon_j\})$ be a pair of reciprocal bases for $V$ (i.e., $\varepsilon^j \cdot \varepsilon_k = \delta^j_k$). We can then construct exactly two natural basis for the space $\bigwedge V$, respectively

$$\{1, \varepsilon^{j_1}, \ldots, \varepsilon^{j_1} \wedge \ldots \wedge \varepsilon^{j_k}, \ldots, \varepsilon^{j_1} \wedge \ldots \wedge \varepsilon^{j_n}\}, \tag{2.12a}$$

and

$$\{1, \varepsilon_{j_1}, \ldots, \varepsilon_{j_1} \wedge \ldots \wedge \varepsilon_{j_k}, \ldots, \varepsilon_{j_1} \wedge \ldots \wedge \varepsilon_{j_n}\}. \tag{2.12b}$$

Then, any $X \in \bigwedge V$ may be expressed using those basis as :

$$X = X \cdot 1 + \sum_{k=1}^{n} \frac{1}{k!} X \cdot (\varepsilon^{j_1} \wedge \ldots \wedge \varepsilon^{j_k})(\varepsilon_{j_1} \wedge \ldots \wedge \varepsilon_{j_k}), \tag{2.13a}$$

$$= X \cdot 1 + \sum_{k=1}^{n} \frac{1}{k!} X \cdot (\varepsilon_{j_1} \wedge \ldots \wedge \varepsilon_{j_k})(\varepsilon^{j_1} \wedge \ldots \wedge \varepsilon^{j_k}), \tag{2.13b}$$

or in a more compact notation, using the collective index[4] $J$,

$$X = \sum_J \frac{1}{\nu(J)!}(X \cdot \varepsilon^J)\varepsilon_J = \sum_J \frac{1}{\nu(J)!}(X \cdot \varepsilon_J)\varepsilon^J. \tag{2.14}$$

## 2.1.4 Canonical Contractions

We now introduce two other important products on $\bigwedge V$, by utilizing only the properties of the canonical scalar product.

Given $X, Y \in \bigwedge V$, the *canonical left contraction* $(X, Y) \mapsto X \lrcorner Y$ is defined by

$$(X \lrcorner Y) \cdot Z := Y \cdot (\widetilde{X} \wedge Z), \tag{2.15}$$

for arbitrary $Z \in \bigwedge V$.

Given $X, Y \in \bigwedge V$, the canonical right contraction $(X, Y) \mapsto X \llcorner Y$ is defined by

$$(X \llcorner Y) \cdot Z := X \cdot (Z \wedge \widetilde{Y}), \tag{2.16}$$

for arbitrary $Z \in \bigwedge V$.

The space $\bigwedge V$, equipped with any of those canonical contractions $\lrcorner$ or $\llcorner$ possesses a natural structure of a non associative algebra. The triple $(\bigwedge V, \lrcorner, \llcorner)$ will be called the canonical interior algebra of multiforms.

Some important properties of $\lrcorner$ and $\llcorner$ are:

**(i)** For all $\alpha, \beta \in \mathbb{R}$ and $X \in \bigwedge^k V$

$$\alpha \lrcorner \beta = \alpha \llcorner \beta = \alpha\beta, \tag{2.17a}$$
$$\alpha \lrcorner X = X \llcorner \alpha = \alpha X, \tag{2.17b}$$

**(ii)** For all $X_j \in \bigwedge^j V$ and $Y_k \in \bigwedge^k V$

$$X_j \lrcorner Y_k = 0, \text{ if } j > k, \tag{2.18}$$
$$X_j \llcorner Y_k = 0, \text{ if } j < k. \tag{2.19}$$

**(iii)** For all $X_j \in \bigwedge^j V$ and $Y_k \in \bigwedge^k V$, with $j \leq k$, we have that $X_j \lrcorner Y_k \in \bigwedge^{k-j} V$, $Y_k \llcorner X_j \in \bigwedge^{k-j} V$, and

$$X_j \lrcorner Y_k = (-1)^{j(k-j)} Y_k \llcorner X_j. \tag{2.20}$$

**(iv)** For all $X_k, Y_k \in \bigwedge^k V$

$$X_k \lrcorner Y_k = X_k \llcorner Y_k = \widetilde{X_k} \cdot Y_k = X_k \cdot \widetilde{Y_k}. \tag{2.21}$$

**(v)** For all $a \in \bigwedge^1 V$ and $X \in \bigwedge V$

$$a \lrcorner X = -\hat{X} \llcorner a. \tag{2.22}$$

---

[4]The collective index $J$ take values over the following set of indices: $\emptyset$ (the null set), $j_1, \ldots, j_1 \ldots j_k, \ldots, j_1 \ldots j_n$. The symbol $\varepsilon^J$ denotes then a scalar basis, $\varepsilon^\emptyset = 1$, a basis of 1-forms $\varepsilon^{j_1}, \ldots$, and $\varepsilon^{j_1 \cdots j_k} = \varepsilon^{j_1} \wedge \ldots \wedge \varepsilon^{j_k}$ abasis for $\bigwedge^k V$. We will also use ,when convenient, the notation $\varepsilon^{j_1 \cdots j_n} = \varepsilon^{j_1} \wedge \ldots \wedge \varepsilon^{j_n}$. Finally, we note that $\nu(J)$ is the number of indices, defined by $\nu(J) = 0, 1, \ldots, k, \ldots, n$ with $J$ taking values over the set just introduced.

**(vi)** For all $a \in \bigwedge^1 V$ and $X, Y \in \bigwedge V$

$$a \lrcorner (X \wedge Y) = (a \lrcorner X) \wedge Y + \hat{X} \wedge (a \lrcorner Y). \tag{2.23}$$

**(vii)** For all $X, Y, Z \in \bigwedge V$

$$X \lrcorner (Y \lrcorner Z) = (X \wedge Y) \lrcorner Z, \tag{2.24a}$$
$$(X \llcorner Y) \llcorner Z = X \llcorner (Y \wedge Z). \tag{2.24b}$$

## 2.2 The Canonical Clifford Algebra

We will define now on $\bigwedge V$ the *canonical Clifford product* (or *geometrical product*) utilizing the canonical contractions and the exterior product. For any $X, Z \in \bigwedge V$ the Clifford product of $X$ by $Y$ will be denoted by juxtaposition of symbols and it obeys the following axioms:
**(i)** For all $\alpha \in \mathbb{R}$ and $X \in \bigwedge V$

$$\alpha X = X\alpha \tag{2.25}$$

**(ii)** For all $a \in \bigwedge^1 V$ and any $X \in \bigwedge V$

$$aX = a \lrcorner X + a \wedge X, \tag{2.26a}$$
$$Xa = X \llcorner a + X \wedge a. \tag{2.26b}$$

**(iii)** For all $X, Y, Z \in \bigwedge V$

$$X(YZ) = (XY)Z. \tag{2.27}$$

The canonical Clifford product is distributive and associative (precisely axiom **(iii)**). Its distributive law follows directly from the corresponding distributive laws of the canonical contractions and of the exterior product.

The linear space $\bigwedge V$, equipped with the canonical Clifford product, is a (real) associative algebra. It will be called here the canonical Clifford algebra of multiforms *and denoted by* $\mathcal{C}\ell(V, \cdot)$.

We will list now the main properties of the canonical Clifford product.

**(i)** For all $X, Y \in \bigwedge V$

$$\widehat{(XY)} = \hat{X}\,\hat{Y}, \tag{2.28a}$$
$$\widetilde{XY} = \tilde{Y}\tilde{X}. \tag{2.28b}$$

**(ii)** For all $a \in \bigwedge^1 V$ and $X \in \bigwedge V$

$$a \lrcorner X = \frac{1}{2}(aX - \hat{X}\,a). \tag{2.29a}$$
$$a \wedge X = \frac{1}{2}(aX + \hat{X}a). \tag{2.29b}$$

**(iii)** For all $X, Y \in \bigwedge V$

$$X \cdot Y = \left\langle \widetilde{X} Y \right\rangle_0 = \left\langle X \widetilde{Y} \right\rangle_0 . \tag{2.30}$$

**(iv)** Let $\tau \in \bigwedge^n V$, $a \in \bigwedge^1 V$ and $X \in \bigwedge V$, then

$$\tau(a \wedge X) = (-1)^{n-1} a \lrcorner (\tau X). \tag{2.31}$$

This formula is sometimes referred to in the mathematical literature as the *duality identity*.

## 2.3 Extensors

### 2.3.1 The Space *extV*

A linear mapping between two arbitrary parts[5] $\bigwedge_1^\diamond V, \bigwedge_2^\diamond V$ of $\bigwedge V$, i.e., $t : \bigwedge_1^\diamond V \to \bigwedge_2^\diamond V$ such that for all $\alpha, \alpha' \in \mathbb{R}$ and $X, X' \in \bigwedge_1^\diamond V$ we have

$$t(\alpha X + \alpha' X') = \alpha t(X) + \alpha' t(X'), \tag{2.32}$$

will be called an extensor over $V$. The set of all extensor over $V$ whose domain and codomain are $\bigwedge_1^\diamond V$ and $\bigwedge_2^\diamond V$ possesses a natural structure of vector space over $\mathbb{R}$.

When $\bigwedge_1^\diamond V = \bigwedge_2^\diamond V = \bigwedge V$, the space of extensors $t : \bigwedge V \to \bigwedge V$ will be denoted by $extV$.

### 2.3.2 The Space $(p, q)$-*extV* of the $(p, q)$-Extensors

Let $p, q$ be two integers with $0 \leq p, q \leq n$. An extensor with domain $\bigwedge^p V$and codomain $\bigwedge^q V$is said to be a $(p, q)$-*extensor over* $V$. The real vector space of the $(p, q)$-extensors will be denoted $(p, q)$-$extV$. If $\dim \mathbf{V} = n$, then $\dim((p, q)\text{-}extV) = \binom{n}{p}\binom{n}{q}$.

### 2.3.3 The Adjoint Operator

Let $(\{\varepsilon^j\}, \{\varepsilon_j\})$ be a pair of arbitrary reciprocal bases for $V$ (i.e., $\varepsilon^j \cdot \varepsilon_i = \delta_i^j$). The linear operator, acting on the space $(p, q)$-$extV \ni t \mapsto t^\dagger \in (q, p)$-$extV$, such that

$$t^\dagger(X) := \frac{1}{p!}(t(\varepsilon^{j_1} \wedge \ldots \wedge \varepsilon^{j_p}) \cdot X)\varepsilon_{j_1} \wedge \ldots \wedge \varepsilon_{j_p} \tag{2.33a}$$

$$= \frac{1}{p!}(t(\varepsilon_{j_1} \wedge \ldots \wedge \varepsilon_{j_p}) \cdot X)\varepsilon^{j_1} \wedge \ldots \wedge \varepsilon^{j_p}, \tag{2.33b}$$

is called the *adjoint operator* and $t^\dagger$ is read as the adjoint of $t$.

The adjoint operator is, of course, well defined since the sums appearing in Eq.(2.33a) and Eq.(2.33b) do not depend on the choice of the pair of reciprocal basis $(\{\varepsilon^j\}, \{\varepsilon_j\})$.

We give now some important properties of the adjoint operator.

---

[5]A direct sum of any vector spaces $\bigwedge^k V$ is clearly a subspace of $\bigwedge V$ and is said to be a *part* of $\bigwedge V$. A convenient notation for a part of $\bigwedge V$ is $\bigwedge^\diamond V$.

**(i)** The adjoint operator is involutive, i.e.,

$$(t^{\dagger})^{\dagger} = t. \tag{2.34}$$

**(ii)** Let $t \in (p,q)$-$extV$. Then, for all $X \in \bigwedge^p V$, $Y \in \bigwedge^q V$ we have that

$$t(X) \cdot Y = X \cdot t^{\dagger}(Y). \tag{2.35}$$

**(iii)** Let $t \in (q,r)$-$extV$ and $u \in (p,q)$-$extV$. The extensor $t \circ u$ (composition of $u$ with $t$), which clearly belongs to $(p,r)$-$extV$, satisfies

$$(t \circ u)^{\dagger} = u^{\dagger} \circ t^{\dagger}. \tag{2.36}$$

### *2.3.4 (1,1)-Extensors, Properties and Associated Extensors*

Symmetric and Antisymmetric parts of $(1,1)$-Extensors

An extensor $t \in (1,1)$-$extV$ is said to be *adjoint symmetric* (respectively *adjoint antisymmetric*) iff $t = t^{\dagger}$ (respectively $t = -t^{\dagger}$).

The following result is important. For any $t \in (1,1)$-$extV$, there exist exactly two $(1,1)$-extensors over $V$, say $t_+$ and $t_-$, such that $t_+$ is adjoint symmetric (i.e., $t_+ = t_+^{\dagger}$) and $t_-$ is adjoint antisymmetric (i.e., $t_- = -t_-^{\dagger}$) and the following decomposition is valid:

$$t(a) = t_+(a) + t_-(a). \tag{2.37}$$

Those $(1,1)$-extensors are given by

$$t_{\pm}(a) = \frac{1}{2}(t(a) \pm t^{\dagger}(a)). \tag{2.38}$$

The extensors $t_+$ and $t_-$ are called the adjoint symmetric and the adjoint antisymmetric parts *of* $t$.

The extension of $(1,1)$-Extensors

Let $(\{\varepsilon^j\}, \{\varepsilon_j\})$ be an arbitrary pair of reciprocal basis for$V$ (i.e., $\varepsilon^j \cdot \varepsilon_i = \delta_i^j$) and let $t \in (1,1)$-$extV$. The *extension operator* is the linear mapping

$$\underline{\ \ } : (1,1)\text{-}extV \to extV,$$
$$t \mapsto \underline{t}, \tag{2.39}$$

such that

$$\underline{t}(X) := 1 \cdot X + \sum_{k=1}^{n} \frac{1}{k!}((\varepsilon^{j_1} \wedge \ldots \wedge \varepsilon^{j_k}) \cdot X) t(\varepsilon_{j_1}) \wedge \ldots \wedge t(\varepsilon_{j_k}) \tag{2.40}$$

$$= 1 \cdot X + \sum_{k=1}^{n} \frac{1}{k!}((\varepsilon_{j_1} \wedge \ldots \wedge \varepsilon_{j_k}) \cdot X) t(\varepsilon^{j_1}) \wedge \ldots \wedge t(\varepsilon^{j_k}). \tag{2.41}$$

In what follows, we say that $\underline{t}$ is the *extension of* $t$ [6].

The extension operator is well defined since the sums appearing in Eq.(2.40) and Eq.(2.41) do not depend on the pair $(\{\varepsilon^j\}, \{\varepsilon_j\})$. Moreover, the extension operator preserves the graduation, i.e., if $X \in \bigwedge^k V$, then $\underline{t}(X) \in \bigwedge^k V$.

We will present now some important properties of the $(1,1)$-extensors.

**(i)** Let $t \in (1,1)\text{-}extV$, for all $\alpha \in \mathbb{R}$, $v \in \bigwedge V$ and $v_1, \ldots, v_k \in \bigwedge^1 V$ we have that

$$\underline{t}(\alpha) = \alpha, \tag{2.42a}$$

$$\underline{t}(v) = t(v), \tag{2.42b}$$

$$\underline{t}(v_1 \wedge \ldots \wedge v_k) = t(v_1) \wedge \ldots \wedge t(v_k). \tag{2.42c}$$

The last property possesses an immediate corollary,

$$\underline{t}(X \wedge Y) = \underline{t}(X) \wedge \underline{t}(Y), \tag{2.43}$$

for all $X, Y \in \bigwedge V$.

**(ii)** For all $t, u \in (1,1)\text{-}extV$, the following identity holds;

$$\underline{t \circ u} = \underline{t} \circ \underline{u}. \tag{2.44}$$

**(iii)** Let $t \in (1,1)\text{-}extV$ with inverse $t^{-1} \in (1,1)\text{-}extV$ (i.e., $t \circ t^{-1} = t^{-1} \circ t = i_d$, where $i_d \in (1,1)\text{-}extV$ is the identity extensor), then

$$(\underline{t})^{-1} = \underline{(t^{-1})}, \tag{2.45}$$

and we shall use the symbol $\underline{t}^{-1}$ to denote both extensors $(\underline{t})^{-1}$ and $\underline{(t^{-1})}$.

**(iv)** For any $(1,1)$-extensor over $V$, the extension operator commutes with the adjoint operator, i.e.,

$$\underline{(t^\dagger)} = (\underline{t})^\dagger, \tag{2.46}$$

and we shall use the symbol $\underline{t}^\dagger$ to denote both extensors $\underline{(t^\dagger)}$ and $(\underline{t})^\dagger$.

**(v)** Let $t \in (1,1)\text{-}extV$, for all $X, Y \in \bigwedge V$ we have

$$X \lrcorner \underline{t}(Y) = \underline{t}(\underline{t}^\dagger(X) \lrcorner Y). \tag{2.47}$$

### The Characteristic Scalars tr[$t$]and det[$t$]

Let $(\{\varepsilon^j\}, \{\varepsilon_j\})$ be an arbitrary pair of reciprocal bases for $V$ (i.e., $\varepsilon^j \cdot \varepsilon_i = \delta_i^j$). The *trace mapping*

$$\begin{aligned} \mathrm{tr} : (1,1)\text{-}extV &\to \mathbb{R}, \\ t &\mapsto tr[t] \end{aligned} \tag{2.48}$$

is such that

$$\mathrm{tr}[t] := t(\varepsilon^j) \cdot \varepsilon_j = t(\varepsilon_j) \cdot \varepsilon^j. \tag{2.49}$$

[6] Some authors call $\underline{t}$ the *exterior algebra extension* of $t$.

Observe that the evaluation of tr does not dependent on the pair $(\{\varepsilon^j\},\{\varepsilon_j\})$ used in Eq.(2.49). We have immediately that, for any $t\in(1,1)$-*extV*,

$$\operatorname{tr}[t^\dagger]=\operatorname{tr}[t]. \tag{1.42}$$

The *determinant mapping*

$$\begin{aligned}\det[t] &: (1,1)\text{-}extV\to\mathbb{R},\\ t &\mapsto \det[t],\end{aligned}$$

is such that

$$\det[t]:=\frac{1}{n!}\underline{t}(\varepsilon^{j_1}\wedge\ldots\wedge\varepsilon^{j_n})\cdot(\varepsilon_{j_1}\wedge\ldots\wedge\varepsilon_{j_n}) \tag{2.50a}$$

$$=\frac{1}{n!}\underline{t}(\varepsilon_{j_1}\wedge\ldots\wedge\varepsilon_{j_n})\cdot(\varepsilon^{j_1}\wedge\ldots\wedge\varepsilon^{j_n}). \tag{2.50b}$$

Like the trace, the determinant also does not depend on the pair of arbitrary reciprocal bases $(\{\varepsilon^j\},\{\varepsilon_j\})$.

By using the combinatorial formulas $v^{j_1}\wedge\ldots\wedge v^{j_n}=\epsilon^{j_1\ldots j_n}v^1\wedge\ldots\wedge v^n$ and $v_{j_1}\wedge\ldots\wedge v_{j_n}=\epsilon_{j_1\ldots j_n}v_1\wedge\ldots\wedge v_n$, where $\epsilon^{j_1\ldots j_n}$ and $\epsilon_{j_1\ldots j_n}$ are symbols of permutation[7] of order $n$ and $v^1,\ldots,v^n$ and $v_1,\ldots,v_n$ are 1-forms, we can write formulas more convenient for $\det t$. Indeed,we have:

$$\det[t]=\underline{t}(\varepsilon^1\wedge\ldots\wedge\varepsilon^n)\cdot(\varepsilon_1\wedge\ldots\wedge\varepsilon_n)=\underline{t}(\overline{\varepsilon})\cdot\underline{\varepsilon} \tag{2.51a}$$

$$=\underline{t}(\varepsilon_1\wedge\ldots\wedge\varepsilon_n)\cdot(\varepsilon^1\wedge\ldots\wedge\varepsilon^n)=\underline{t}(\underline{\varepsilon})\cdot\overline{\varepsilon}, \tag{2.51b}$$

where we used the short notations: $\overline{\varepsilon}=\varepsilon^1\wedge\ldots\wedge\varepsilon^n$ and $\underline{\varepsilon}=\varepsilon_1\wedge\ldots\wedge\varepsilon_n$.

The mapping det possesses the following important properties.

**(i)** For all $t\in(1,1)$-*extV*,

$$\det[t^\dagger]=\det[t]. \tag{2.52}$$

**(ii)** Let $t\in(1,1)$-*extV*, for any $\tau\in\bigwedge V$ we have

$$\underline{t}(\tau)=\det[t]\tau. \tag{2.53}$$

**(iii)** For all $t,u\in(1,1)$-*extV*,

$$\det[t\circ u]=\det[t]\circ\det[u]. \tag{2.54}$$

**(iv)** Let $t\in(1,1)$-*extV* with inverse $t^{-1}\in(1,1)$-*extV* (i.e., $t\circ t^{-1}=t^{-1}\circ t=i_d$, where $i_d\in(1,1)$-*extV* is the identity extensor). Then,

$$\det[t^{-1}]=(\det[t])^{-1}. \tag{2.55}$$

We shall use the notation $\det^{-1}[t]$ to denote both $\det[t^{-1}]$ and $(\det[t])^{-1}$.

**(v)** If $t\in(1,1)$-*extV* is non degenerate (i.e., $\det[t]\neq0$), then its inverse $t^{-1}\in(1,1)$-*extV* exists, and is given by

$$t^{-1}(a)=\det{}^{-1}[t]\underline{t}^\dagger(a\tau)\tau^{-1}, \tag{2.56}$$

where $a\in\bigwedge^1V$ and $\tau\in\bigwedge^nV$, $\tau\neq0$.

---

[7] The permutation symbols of order $n$ are defined by

$$\epsilon^{j_1\ldots j_n}=\epsilon_{j_1\ldots j_n}=\begin{cases}1, & \text{if } j_1\ldots j_n \text{ is an even permutation of } 1,\ldots,n\\ -1, & \text{if } j_1\ldots j_n \text{ is an odd permutation of } 1,\ldots,n\\ 0, & \text{in all other cases.}\end{cases}.$$

Moreover, recall that $\epsilon^{j_1\ldots j_n}\epsilon_{j_1\ldots j_n}=n!$

### The Characteristic Biform Mapping bif

Let $(\{\varepsilon^j\},\{\varepsilon_j\})$ be an arbitrary pair of reciprocal bases for $V$ (i.e., $\varepsilon^j\cdot\varepsilon_i=\delta_i^j$). The *biform mapping*

$$\text{bif}:(1,1)\text{-}extV\to\bigwedge^2 V,$$

is such that

$$\text{bif}[t]:=t(\varepsilon^j)\wedge\varepsilon_j=t(\varepsilon_j)\wedge\varepsilon^j. \tag{2.57}$$

Note that $\text{bif}[t]$ is a 2-form characteristic of the extensor $t$ (since it does not depend on the pair of reciprocal bases $(\{\varepsilon^j\},\{\varepsilon_j\})$, and $\text{bif}[t]$ is read as the *biform* of $t$.

We will give now some important properties of bif mapping:

**(i)** Let $t\in(1,1)\text{-}extV$, then

$$\text{bif}[t^\dagger]=-\text{bif}[t]. \tag{2.58}$$

**(ii)** The antisymmetric adjoint of any $t\in(1,1)\text{-}extV$ may be factored by the following formula

$$t_-(a)=\frac{1}{2}\text{bif}[t]\times a, \tag{2.59}$$

where $\times$ means here the canonical commutator defined for any $A,B\in\bigwedge V$ by

$$A\times B:=\frac{1}{2}(AB-BA). \tag{2.60}$$

### The Generalization Operator of $(1,1)$-Extensors

Let $(\{\varepsilon^j\},\{\varepsilon_j\})$ be an arbitrary pair of reciprocal basis for $V$ (i.e., $\varepsilon^j\cdot\varepsilon_i=\delta_i^j$). The linear operator whose domain is the space $(1,1)\text{-}extV$ and whose codomain is the space $extV$, given by

$$\begin{gathered}(1,1)\text{-}extV\ni t\mapsto T\in extV,\\ T(X)=t(\varepsilon^j)\wedge(\varepsilon_j\lrcorner X)=t(\varepsilon_j)\wedge(\varepsilon^j\lrcorner X),\end{gathered} \tag{2.61}$$

is called the *generalization operator*, and $T$ is read as the *generalized of* $t$.

The generalization operator is well defined since the sum in Eq.(2.61) does not depend on $(\{\varepsilon^j\},\{\varepsilon_j\})$. The generalization operators preserves the grade, i.e., if $X\in\bigwedge^k V$, then $T(X)\in\bigwedge^k V$.

We will list now some important properties of the generalization of a (1,1)-extensor $t$:

**(i)** For all $\alpha\in\mathbb{R}$ and $v\in\bigwedge^1 V$,

$$T(\alpha)=0, \tag{2.62a}$$
$$T(v)=t(v). \tag{2.62b}$$

**(ii)** For all $X,Y\in\bigwedge V$, it is:

$$T(X\wedge Y)=T(X)\wedge Y+X\wedge T(Y). \tag{2.63}$$

**(iii)** The generalization operator commutes with the adjoint operator. Then, $T^\dagger$ denotes both the adjoint of the generalized and the generalized of the *adjoint.*

**(iv)** The antisymmetric part of the adjoint of the generalized coincides with the generalized of the adjoint antisymmetric part of $t$ and can then be factored as

$$T_-(X)=\frac{1}{2}\text{bif}[t]\times X, \tag{2.64}$$

for all $X\in\bigwedge V$.

Normal $(1,1)$-Extensors

An extensor $t \in (1,1)\text{-}extV$ is said to be *normal* if its adjoint symmetric and antisymmetric parts commute, i.e., for any $a \in V$,

$$t_+(t_-(a)) = t_-(t_+(a)). \tag{2.65}$$

We can show, without difficulties [3], that Eq.(2.65) is equivalent to any of the following two equations

$$\begin{aligned} \underline{t}(\underline{t}^\dagger(A)) &= \underline{t}^\dagger(\underline{t}(A)),\\ \langle \underline{t}(A)\underline{t}(B)\rangle_0 &= \langle \underline{t}^\dagger(A)\underline{t}^\dagger(B)\rangle_0, \end{aligned} \tag{2.66}$$

for any $A, B \in \bigwedge V$.

## 2.4 The Metric Clifford Algebra $\mathcal{C}\ell(V, \boldsymbol{g})$

A $(1,1)$-extensor over $V$, say $\boldsymbol{g}$, which is symmetric adjoint (i.e., $\boldsymbol{g} = \boldsymbol{g}^\dagger$) and non degenerate (i.e., $\det \boldsymbol{g} \neq 0$), is said to be a metric extensor on $V$. Under these conditions, $\boldsymbol{g}^{-1}$ exists and is also a metric extensor on $V$, and as we are going to see, both $\boldsymbol{g}$ and $\boldsymbol{g}^{-1}$ have equally important roles in our theory. The reason is that we are going to use $\boldsymbol{g}$ to represent the metric tensor $\boldsymbol{g} \in T_2^0\mathbf{V}$, whereas $\boldsymbol{g}^{-1}$ will represent $\mathtt{g} \in T_0^2\mathbf{V}$. This is done as follows. Given an arbitrary basis $\{\mathbf{e}_i\}$ of $\mathbf{V}$ and the corresponding dual basis $\{\varepsilon^i\}$ de $V$ and the reciprocal basis of $\{\varepsilon^i\}$, i.e., $\varepsilon_i \cdot \varepsilon^j = \delta_i^j$, if $\mathbf{g}(\mathbf{e}_i, \mathbf{e}_j) = g_{ij}$ and $\mathtt{g}(\varepsilon^i, \varepsilon^j) = g^{ij}$ with $g_{ij}g^{jk} = \delta_i^k$, then:

$$\begin{aligned} \boldsymbol{g}^{-1}(\varepsilon^i) \cdot \varepsilon^j &= g^{ij},\\ \boldsymbol{g}(\varepsilon_i) \cdot \varepsilon_j &= g_{ij}. \end{aligned} \tag{2.67}$$

The Metric Scalar Product

Given a metric extensor $\boldsymbol{g} \in (1,1)\text{-}extV$, we may define two new scalar products on $\bigwedge V$, respectively $\underset{g}{\cdot}$ and $\underset{g^{-1}}{\cdot}$ such that:

$$X \underset{g}{\cdot} Y := \underline{\boldsymbol{g}}(X) \cdot Y, \tag{2.68}$$

$$X \underset{g^{-1}}{\cdot} Y := \underline{\boldsymbol{g}}^{-1}(X) \cdot Y \tag{2.69}$$

Both $\underset{g}{\cdot}$ and $\underset{g^{-1}}{\cdot}$ will be called a *metric scalar product.*

The mappings $\underset{g}{\cdot}$ and $\underset{g^{-1}}{\cdot}$ are indeed well defined scalar products on $\bigwedge V$. The scalar product $\underset{g}{\cdot}$ (and obviously also $\underset{g^{-1}}{\cdot}$) is symmetric (i.e., $X \underset{g}{\cdot} Y = Y \underset{g}{\cdot} X$), is linear (i.e., $X \underset{g}{\cdot} (Y + Z) = X \underset{g}{\cdot} Y + X \underset{g}{\cdot} Z$) Moreover, it also satisfies $(\alpha X) \underset{g}{\cdot} Y = X \underset{g}{\cdot} (\alpha Y) = \alpha(X \underset{g}{\cdot} Y)$, $\forall \alpha \in \mathbb{R}$, and is non degenerate (i.e.,

$X \underset{g}{\cdot} Y = 0$ for all $X$, implies $Y = 0$). In what follows, we will list the main properties of $\underset{g}{\cdot}$, since $\underset{g^{-1}}{\cdot}$ obey similar properties.

**(i)** For $\alpha, \beta \in \mathbb{R}$

$$\alpha \underset{g}{\cdot} \beta = \alpha\beta, \tag{2.70}$$

**(ii)** For all $X_j \in \bigwedge^j V$ and $Y_k \in \bigwedge^k V$

$$X_j \underset{g}{\cdot} Y_k = 0, \text{ if } j \neq k. \tag{2.71}$$

**(iii)** For all simple $k$-forms $v_1 \wedge \ldots \wedge v_k \in \bigwedge^k V$ and $w_1 \wedge \ldots \wedge w_k \in \bigwedge^k V$, we have

$$(v_1 \wedge \ldots \wedge v_k) \underset{g}{\cdot} (w_1 \wedge \ldots \wedge w_k) = \begin{vmatrix} v_1 \underset{g}{\cdot} w_1 & \ldots & v_1 \underset{g}{\cdot} w_k \\ \vdots & & \vdots \\ v_k \underset{g}{\cdot} w_1 & \ldots & v_k \underset{g}{\cdot} w_k \end{vmatrix}, \tag{2.72}$$

where, as before, $|\ |$ denotes the classical determinant of the matrix with the entries $v_i \underset{g}{\cdot} w_j$ .

**(iv)** For all $X, Y \in \bigwedge V$

$$\hat{X} \underset{g}{\cdot} Y = X \underset{g}{\cdot} \hat{Y}. \tag{2.73a}$$

$$\widetilde{X} \underset{g}{\cdot} Y = X \underset{g}{\cdot} \widetilde{Y}. \tag{2.73b}$$

The Metric Left and Right Contractions

The metric left and right contractions of $X, Y \in \bigwedge V$ are given by

$$X \underset{g}{\lrcorner} Y := \underline{\boldsymbol{g}}(X) \lrcorner Y, \tag{2.74}$$

$$X \underset{g}{\llcorner} Y := X \llcorner \underline{\boldsymbol{g}}(Y). \tag{2.75}$$

The linear space $\bigwedge V$, equipped with any one of those contractions, has a structure of non associative algebra, and $(\bigwedge V, \underset{g}{\lrcorner}, \underset{g}{\llcorner})$ will be called metric interior algebra of multiforms.

The most important properties of the metric left and right contractions are:

**(i)** For all $X, Y, Z \in \bigwedge V$,

$$(X \underset{g}{\lrcorner} Y) \underset{g}{\cdot} Z = Y \underset{g}{\cdot} (\widetilde{X} \wedge Z), \tag{2.76a}$$

$$(X \underset{g}{\llcorner} Y) \underset{g}{\cdot} Z = X \underset{g}{\cdot} (Z \wedge \widetilde{Y}). \tag{2.76b}$$

**(ii)** For all $\alpha, \beta \in \mathbb{R}$ and $X \in \bigwedge V$

$$\alpha \underset{g}{\lrcorner} \beta = \alpha \underset{g}{\llcorner} \beta = \alpha\beta, \tag{2.77a}$$

$$\alpha \underset{g}{\lrcorner} X = X \underset{g}{\llcorner} \alpha = \alpha X, \tag{2.77b}$$

**(iii)** For all $X_j \in \bigwedge^j V$ and $Y_k \in \bigwedge^k V$

$$X_j \underset{g}{\lrcorner} Y_k = 0, \text{ if } j > k. \tag{2.78a}$$

$$X_j \underset{g}{\llcorner} Y_k = 0, \text{ if } j < k. \tag{2.78b}$$

**(iv)** For all $X_j \in \bigwedge^j V$ and $Y_k \in \bigwedge^k V$, with $j \leq k$, we have: $X_j \underset{g}{\lrcorner} Y_k \in \bigwedge^{k-j} V$ and $Y_k \underset{g}{\llcorner} X_j \in \bigwedge^{k-j} V$, and

$$X_j \underset{g}{\lrcorner} Y_k = (-1)^{j(k-j)} Y_k \underset{g}{\llcorner} X_j. \tag{2.79}$$

**(v)** For all $X_k, Y_k \in \bigwedge^k V$

$$X_k \underset{g}{\lrcorner} Y_k = X_k \underset{g}{\llcorner} Y_k = \widetilde{X}_k \underset{g}{\cdot} Y_k = X_k \underset{g}{\cdot} \widetilde{Y}_k. \tag{2.80}$$

**(vi)** For all $a \in \bigwedge^1 V$ and $X \in \bigwedge V$

$$a \underset{g}{\lrcorner} X = -\hat{X} \underset{g}{\llcorner} a. \tag{2.81}$$

**(vii)** For all $a \in \bigwedge^1 V$ and $X, Y \in \bigwedge V$

$$a \underset{g}{\lrcorner} (X \wedge Y) = (a \underset{g}{\lrcorner} X) \wedge Y + \hat{X} \wedge (a \underset{g}{\lrcorner} Y). \tag{2.82}$$

The Metric Clifford Product

The metric Clifford product of two arbitrary elements $X, Y \in \bigwedge V$ will be written $X \underset{g}{\ } Y$, and it satisfies the following axioms:

**(i)** For all $\alpha \in \mathbb{R}$ and $X \in \bigwedge V$

$$\alpha \underset{g}{\ } X = X \underset{g}{\ } \alpha = \alpha X,$$

**(ii)** For all $a \in \bigwedge^1 V$ and $X \in \bigwedge V$

$$a \underset{g}{\ } X = a \underset{g}{\lrcorner} X + a \wedge X,$$

$$X \underset{g}{\ } a = X \underset{g}{\llcorner} a + X \wedge a.$$

**(iii)** For all $X, Y, Z \in \bigwedge V$

$$X \underset{g}{\ } (Y \underset{g}{\ } Z) = (X \underset{g}{\ } Y) \underset{g}{\ } Z.$$

The metric Clifford product is distributive and associative. Its distributive law follows from the corresponding distributive laws satisfied by the left and right contractions and the exterior product. Its associative law is just the axiom **(iii)**.

$\bigwedge V$ equipped with a metric Clifford product is an associative algebra, called a *metric Clifford algebra over* $V$. It will be denoted by $\mathcal{C}\ell(V, \boldsymbol{g})$. In a totally similar way we may define the algebra $\mathcal{C}\ell(V, \boldsymbol{g}^{-1})$.

We will conclude this section by listing some useful properties of the metric Clifford product, which will be used in the calculations of the following chapters.

**(i)** For all $a \in \bigwedge^1 V$ and $X, Y, Z \in \bigwedge V$ :

$$\begin{aligned} \widehat{X \underset{g}{} Y} &= \hat{X} \underset{g}{} \hat{Y}, \\ \widetilde{X \underset{g}{} Y} &= \widetilde{X} \underset{g}{} \widetilde{Y}. \\ a \underset{g}{\lrcorner} X &= \frac{1}{2}(a \underset{g}{} X - \hat{X} \underset{g}{} a), \\ a \wedge X &= \frac{1}{2}(a \underset{g}{} X + \hat{X} \underset{g}{} a). \end{aligned} \tag{2.83}$$

$$\begin{aligned} X \underset{g}{\cdot} Y &= \left\langle \widetilde{X} \underset{g}{} Y \right\rangle_0 = \left\langle X \underset{g}{} \widetilde{Y} \right\rangle_0 . \\ X \underset{g}{\lrcorner} (Y \underset{g}{\lrcorner} Z) &= (X \wedge Y) \underset{g}{\lrcorner} Z, \\ (X \underset{g}{\llcorner} Y) \underset{g}{\llcorner} Z &= X \underset{g}{\llcorner} (Y \wedge Z). \end{aligned} \tag{2.84}$$

**(ii)** Let $\tau \in \bigwedge^n V$ then for all $a \in \bigwedge^1 V$ and $X \in \bigwedge V$ :

$$\tau \underset{g}{} (a \wedge X) = (-1)^{n-1} a \underset{g}{\lrcorner} (\tau \underset{g}{} X). \tag{2.85}$$

Eq.(2.85) will be called *metric dual identity.*

## 2.5 Pseudo-Euclidean Metric Extensors on $V$

### 2.5.1 The metric extensor $\eta$

Consider the vector spaces $\mathbf{V}$and $V$ ($\dim \mathbf{V} = \dim V = n$) and the dual bases $\{\mathbf{b}_\mu\}$ and $\{\boldsymbol{\beta}^\mu\}$ for $\mathbf{V}$ and $V$, $\boldsymbol{\beta}^\mu(\mathbf{b}_\nu) = \delta^\mu_\nu$. Moreover, denote the reciprocal basis of the basis $\{\boldsymbol{\beta}^\mu\}$ by $\{\beta_\mu\}$, with $\boldsymbol{\beta}^\mu \cdot \beta_\nu = \delta^\mu_\nu$, and $\mu, \nu = 0, 1, 2..., n$.

The extensor on $V$, say $\eta \in (1,1)$-$extV$ defined for any $a \in \bigwedge^1 V$by $a \mapsto \eta(a) \in \bigwedge^1 V$, such that

$$\eta(a) = \boldsymbol{\beta}^0 a \boldsymbol{\beta}^0, \tag{2.86}$$

is a well defined metric extensor on $V$ (i.e., $\eta$ is symmetric and non degenerate) and it is called a *pseudo-Euclidean metric extensor for* $V$.

The main properties of $\eta$ are:

**(i)** $\boldsymbol{\beta}^0$ is an eigenvector with eigenvalue 1, i.e., $\eta(\boldsymbol{\beta}^0) = \boldsymbol{\beta}^0$, and $\boldsymbol{\beta}^k$ ($k = 1, 2, ..., n$) are eigenvectors with eigenvalues $-1$, i.e., $\eta(\boldsymbol{\beta}^k) = -\boldsymbol{\beta}^k$.

**(ii)** The 1-forms of the canonical basis $\{\boldsymbol{\beta}^\mu\}$ are $\eta$ orthonormal, i.e.,

$$\boldsymbol{\beta}^\mu \underset{\eta}{\cdot} \boldsymbol{\beta}^\nu := \eta(\boldsymbol{\beta}^\mu) \cdot \boldsymbol{\beta}^\nu = \eta^{\mu\nu}, \tag{2.87}$$

where the matrix with entries $(\eta^{\mu\nu})$ is the diagonal matrix $(1, -1, ..., -1)$.

Then, $\eta$ has signature[8] $(1, n)$.

**(iii)** $\mathrm{tr}[\eta] = 2 - n$ and $\det[\eta] = (-1)^{n-1}$.

**(iv)** The extended $\underline{\eta}$ of $\eta$ has the same generator as $\eta$, i.e., for any $X \in \bigwedge U$, $\underline{\eta}(X) = \boldsymbol{\beta}^0 X \boldsymbol{\beta}^0$.

**(v)** $\eta$ is orthogonal canonical, i.e., since

$$\eta(\boldsymbol{\beta}^\mu) \cdot \boldsymbol{\beta}^\nu = \boldsymbol{\beta}^\mu \cdot \eta(\boldsymbol{\beta}^\nu),$$

it is $\eta = \eta^\clubsuit$ (where $\eta^\clubsuit = (\eta^\dagger)^{-1} = (\eta^{-1})^\dagger$). Then, $\eta^2 = i_d$.

**Remark 2.1** When $\dim \mathbf{V} = 4$ and the signature of $\eta$ is $(1, 3)$, the pair $(\mathbf{V},\eta)$ is called a *Minkowski vector space* and $\eta$ is called a Minkowski metric extensor.

### *2.5.2 Metric Extensor* $\boldsymbol{g}$ *with the Same Signature of* $\eta$

Let $\boldsymbol{g}$ be a general metric extensor on $V$with the same signature as the $\eta$ extensor defined by Eq.(2.86). Then, we have the following fundamental theorem[9], which will play a crucial role in the considerations that follow.

**Theorem 2.1** For any pseudo-Euclidean metric extensor $\boldsymbol{g}$ for $V$, there exists an invertible $(1,1)$-extensor $\boldsymbol{h}$ (non unique!), such that

$$\boldsymbol{g} = \boldsymbol{h}^\dagger \eta \boldsymbol{h}, \tag{2.88}$$

where $\eta$ is defined by Eq.(2.86). The $(1,1)$-extensor field $\boldsymbol{h}$ is given by:

$$\boldsymbol{h}(a) = \sum_{\mu=0}^{n} \sqrt{|\lambda^\mu|}(a \cdot v^\mu)\boldsymbol{\beta}^\mu, \tag{2.89}$$

where the $\lambda^\mu$ are the eigenvalues of $\boldsymbol{g}$ and $v^\mu \in \bigwedge^1 V$ are the associated eigenvectors. The 1-form $\{v^\mu\}$ defines a *Euclidean orthonormal basis* for $V$, i.e., $v^\mu \cdot v^\nu = \delta^{\mu\nu}$.

**Proof** We must prove that $\boldsymbol{h}$, in Eq.(2.89), satisfies the equation for the composition of extensors $\boldsymbol{h}^\dagger \eta \boldsymbol{h} = \boldsymbol{g}$.

First, we need to calculate the adjoint of $\boldsymbol{h}$. Take two arbitrary elements $a, b \in \bigwedge^1 V$. Then, using Eq.(2.35) and Eq.(2.89), we can write

$$\begin{aligned} \boldsymbol{h}^\dagger(a) \cdot b &= a \cdot \boldsymbol{h}(b) \\ &= a \cdot (\sum_{\mu=0}^{n} \sqrt{|\lambda^\mu|}(b \cdot v^\mu)\boldsymbol{\beta}^\mu) \\ &= (\sum_{\mu=0}^{n} \sqrt{|\lambda^\mu|}(a \cdot \boldsymbol{\beta}^\mu)v^\mu) \cdot b, \end{aligned} \tag{2.90}$$

[8] Or, as some mathematicians say, signature $(1 - n)$.

[9] Originally proved in [4].

and from the non degeneracy of the scalar product, we have

$$\boldsymbol{h}^{\dagger}(a) = \sum_{\mu=0}^{n} \sqrt{|\lambda^{\mu}|}(a \cdot \boldsymbol{\beta}^{\mu})v^{\mu}. \tag{2.91}$$

Now, using Eqs.(2.89) and (2.91), we get

$$\begin{aligned}
\boldsymbol{h}^{\dagger}\eta\boldsymbol{h}(a) &= \sum_{\mu=0}^{n}\sum_{\nu=0}^{n} \sqrt{|\lambda^{\mu}\lambda^{\nu}|}\eta(\boldsymbol{\beta}^{\mu}) \cdot \boldsymbol{\beta}^{\nu}(a \cdot v^{\mu})v^{\nu} \\
&= \sqrt{|\lambda^{0}|^{2}}\eta(\boldsymbol{\beta}^{0}) \cdot \boldsymbol{\beta}^{0}(a \cdot v^{0})v^{0} + \sum_{j=1}^{n} \sqrt{|\lambda^{j}\lambda^{0}|}\eta(\boldsymbol{\beta}^{j}) \cdot \boldsymbol{\beta}^{0}(a \cdot v^{j})v^{0} \\
&+ \sum_{k=1}^{n} \sqrt{|\lambda^{0}\lambda^{k}|}\eta(\boldsymbol{\beta}^{0}) \cdot \boldsymbol{\beta}^{k}(a \cdot v^{0})v^{k} \\
&+ \sum_{j=1}^{n}\sum_{k=1}^{n} \sqrt{|\lambda^{j}\lambda^{k}|}\eta(\boldsymbol{\beta}^{j}) \cdot \boldsymbol{\beta}^{k}(a \cdot v^{j})v^{k},
\end{aligned}$$

and, since $\eta(\boldsymbol{\beta}^{0}) = \boldsymbol{\beta}^{0}$ and $\eta(\boldsymbol{\beta}^{k}) = -\boldsymbol{\beta}^{k}$ $(k = 1, 2, ..., n)$, and $\boldsymbol{\beta}^{\mu} \cdot \boldsymbol{\beta}^{\nu} = \delta^{\mu\nu}$, we obtain

$$\boldsymbol{h}^{\dagger}\eta\boldsymbol{h}(a) = \left|\lambda^{0}\right|(a \cdot v^{0})v^{0} - \sum_{j=1}^{n}\sum_{k=1}^{n} \sqrt{|\lambda^{j}\lambda^{k}|\delta^{jk}}(a \cdot v^{j})v^{k}$$

$$\boldsymbol{h}^{\dagger}\eta\boldsymbol{h}(a) = \left|\lambda^{0}\right|(a \cdot v^{0})v^{0} - \sum_{j=1}^{n} \left|\lambda^{j}\right|(a \cdot v^{j})v^{j}. \tag{2.92}$$

Finally, using Eqs.(2.91) and (2.90) in Eq.(2.92), we have

$$\begin{aligned}
\boldsymbol{h}^{\dagger}\eta\boldsymbol{h}(a) &= (a \cdot v^{0})\boldsymbol{g}(v^{0}) + \sum_{j=1}^{n}(a \cdot v^{j})\boldsymbol{g}(v^{j}) \\
&= \sum_{\mu=0}^{n}(a \cdot v^{\mu})\boldsymbol{g}(v^{\mu}) = \boldsymbol{g}(\sum_{\mu=0}^{n}(a \cdot v^{\mu})v^{\mu}) \\
\boldsymbol{h}^{\dagger}\eta\boldsymbol{h}(a) &= \boldsymbol{g}(a). \blacksquare
\end{aligned}$$

The $(1,1)$-extensor $\boldsymbol{h}$ given by Eq.(2.89) is invertible since

$$\begin{aligned}
\boldsymbol{h}^{\dagger}\eta\boldsymbol{h} = \boldsymbol{g} &\Rightarrow \det[\boldsymbol{h}^{\dagger}]\det[\eta]\det[\boldsymbol{h}] = \det[\boldsymbol{g}] \\
&\Rightarrow (\det[\boldsymbol{h}])^{2}(-1) = \det[\boldsymbol{g}],
\end{aligned} \tag{2.93}$$

and this implies that $\det[\boldsymbol{h}] \neq 0$, i.e., $\boldsymbol{h}$ is non degenerate and thus invertible.

On the other hand, the $(1,1)$-extensor $\boldsymbol{h}$, which satisfies Eq.(2.88) is certainly not unique. There is a *gauge freedom*, which is represented by a local pseudo orthogonal transformation. Indeed, consider the $(1,1)$-extensor $\boldsymbol{h}' \equiv \boldsymbol{l}\boldsymbol{h}$, where $\boldsymbol{l}$ is an $\eta$-orthogonal $(1,1)$-extensor $\boldsymbol{l} = \boldsymbol{l}^{(\eta)}$, (or equivalently,

$\boldsymbol{l}^{\dagger}\eta\boldsymbol{l} = \eta$) [10] and $\boldsymbol{h}$ is any $(1,1)$-extensor that satisfies Eq.(2.88) (e.g., the $\boldsymbol{h}$ given by Eq.(2.89)), then

$$(\boldsymbol{h}')^{\dagger}\eta\boldsymbol{h}' = (\boldsymbol{l}\boldsymbol{h})^{\dagger}\eta\boldsymbol{l}\boldsymbol{h} = \boldsymbol{h}^{\dagger}l^{\dagger}\eta\boldsymbol{l}\boldsymbol{h} = \boldsymbol{h}^{\dagger}\eta\boldsymbol{h} = \boldsymbol{g}.$$

i.e., $\boldsymbol{h}'$ also satisfies Eq.(2.88).

**Remark 2.2** When $\dim V = 4$ and the signature of $\boldsymbol{g}$ (and, of course, of $\eta$) is $(1,3)$, $\boldsymbol{g}$ is said a *Lorentz metric extensor.*

### *2.5.3 Some Remarkable Results*

We will recall now some remarkable results that will be needed repeatedly in the following chapters.

Golden Rule

Let $\boldsymbol{g} = \boldsymbol{h}^{\dagger}\eta\boldsymbol{h}$ be defined as in Eq.(2.88). Then, for any $X, Y \in \bigwedge U$, it holds the following remarkable identity known as the *golden rule* [5]:

$$\underline{\boldsymbol{h}}(X \underset{g}{*} Y) = \underline{\boldsymbol{h}}(X) \underset{\eta}{*} \underline{\boldsymbol{h}}(Y), \tag{2.94}$$

where $*$ denotes here, as previously agreed, any one of the multiform products.

Hodge Star Operators

The canonical Hodge star operator in $\bigwedge V$ is defined first by its action on $\bigwedge^p V$ by

$$\begin{aligned} \star : \bigwedge^p V &\to \bigwedge^{n-p} V, \\ X \mapsto \star X &:= \tilde{X} \lrcorner \tau = \tilde{X}\tau, \end{aligned} \tag{2.95}$$

where $\tau$ is the canonical volume element[11] $\tau = \boldsymbol{\beta}^0 \wedge \boldsymbol{\beta}^1 \wedge \boldsymbol{\beta}^2 \wedge \ldots \wedge \boldsymbol{\beta}^n$ and next extended by linearity to all $\bigwedge V$.

Then, the Hodge star operators associated with the metric extensors $\eta$ and $\boldsymbol{g}$ just introduced above are given by:

$$\begin{aligned} &\underset{\eta}{\star} : \bigwedge^p V \to \bigwedge^{n-p} V, \quad &&\underset{g}{\star} : \bigwedge^p V \to \bigwedge^{n-p} V, \\ &\underset{\eta}{\star} X := \tilde{X} \lrcorner \tau_\eta = \tilde{X}\tau_\eta, &&\underset{g}{\star} X := \tilde{X} \lrcorner \tau_g = \tilde{X}\tau_g, \end{aligned} \tag{2.96}$$

where

$$\tau_\eta = \sqrt{|\det \eta|}\tau, \; \tau_g = \sqrt{|\det \boldsymbol{g}|}\tau. \tag{2.97}$$

[10] I.e., a *transformation* $l(a) \underset{\eta}{\cdot} l(b) = a \underset{\eta}{\cdot} b$.

[11] Note that if the basis $\{\boldsymbol{\beta}^{\mu}\}$ is (canonical) orthonormal we can write the volume element as $\tau = \boldsymbol{\beta}^0\boldsymbol{\beta}^1\boldsymbol{\beta}^2 \ldots \boldsymbol{\beta}^n$.

Relation Between the Hodge Star Operators of $\boldsymbol{g}$ and the Canonical Hodge Star Operator

We then have the following nontrivial results relating $\underset{g}{\star}$ and $\star$. For any $X \in \bigwedge V$,

$$\underset{g}{\star}X = \frac{1}{\sqrt{|\det \boldsymbol{g}|}}\underline{\boldsymbol{g}}(\tilde{X} \lrcorner \tau) = \frac{1}{\sqrt{|\det \boldsymbol{g}|}}\underline{\boldsymbol{g}} \circ \star X, \tag{2.98}$$

Relation Between the Hodge Star Operators of $\boldsymbol{g}$ and $\eta$

Putting

$$\boldsymbol{h}^{\clubsuit} := \boldsymbol{h}^{\dagger -1}, \tag{2.99}$$

the following result is valid when $\boldsymbol{g}$ and $\eta$ have the same signature

$$\underset{g}{\star} = \operatorname{sgn}(\det \boldsymbol{h})\underline{\boldsymbol{h}}^{\dagger} \circ \underset{\eta}{\star} \circ \underline{\boldsymbol{h}}^{\clubsuit}. \tag{2.100}$$

Moreover, if we suppose that $\boldsymbol{h}$ is continuously connected to the identity extensor which has determinant equal to 1, we can write

$$\underset{g}{\star} = \underline{\boldsymbol{h}}^{\dagger} \underset{\eta}{\star} \underline{\boldsymbol{h}}^{\clubsuit} \tag{2.101}$$

### *2.5.4 Useful Identities*

We shall list some identities involving contractions, exterior product and the Hodge star operator that will be needed for some derivations in Chapters 5, 6,7 and Appendices D and E. These are

For any metric extensor $\boldsymbol{g}$ and for any $A_r \in \bigwedge^r V$ and $B_s \in \bigwedge^s V$, $r, s \geq 0$:

$$\begin{aligned}
&A_r \wedge \underset{g}{\star} B_s = B_s \wedge \underset{g}{\star} A_r; \quad r = s \\
&A_r \underset{g^{-1}}{\cdot} \underset{g}{\star} B_s = B_s \underset{g^{-1}}{\cdot} \underset{g}{\star} A_r; \quad r + s = n \\
&A_r \wedge \underset{g}{\star} B_s = (-1)^{r(s-1)} \underset{g}{\star}(\tilde{A}_r \underset{g^{-1}}{\lrcorner} B_s); \quad r \leq s \\
&A_r \underset{g^{-1}}{\lrcorner} \underset{g}{\star} B_s = (-1)^{rs} \underset{g}{\star}(\tilde{A}_r \wedge B_s); \quad r + s \leq n \\
&\underset{g}{\star}A_r = \tilde{A}_r \underset{g^{-1}}{\lrcorner} \tau_{\underset{g}{\circ}} = \tilde{A}_r \, \tau_g \\
&\underset{g}{\star}\tau_g = \operatorname{sgn}\boldsymbol{g}; \quad \underset{g}{\star}1 = \tau_g.
\end{aligned} \tag{2.102}$$

Finally, we shall also need the following result. Let $\{\varepsilon^\mu\}$ be an arbitrary basis of $U$, such that $\varepsilon^\mu \underset{g^{-1}}{\cdot} \varepsilon^\mu = g^{\mu\nu}$. Then, if

$$\varepsilon^{\mu_1 \ldots \mu_p} = \varepsilon^{\mu_1} \wedge \ldots \wedge \varepsilon^{\mu_p}, \varepsilon^{\nu_{p+1} \ldots \nu_n} = \varepsilon^{\nu_{p+1}} \wedge \ldots \wedge \varepsilon^{\nu_n}$$

we have, with $g^{\mu\nu} g_{\mu\alpha} = \delta^\nu_\alpha$,

$$\underset{g}{\star}\varepsilon^{\mu_1 \ldots \mu_p} = \frac{1}{(n-p)!}\sqrt{|\det(g_{\mu\nu})|}g^{\mu_1\nu_1} \ldots g^{\mu_p\nu_p}\varepsilon_{\nu_1 \ldots \nu_n}\varepsilon^{\nu_{p+1} \ldots \nu_n}, \tag{2.103}$$

where $\det(g_{\mu\nu})$ is the classical determinant of the matrix with entries $g_{\mu\nu}$.

## References

[1] da Rocha, R., and Rodrigues, W. A. Jr., Pair and Impar, Even and Odd Form Fields and Electromagnetism, *Ann. Phys.* (Berlin) **19**, 6-34 (2010). [arXiv:0811.1713]

[2] Rodrigues, W. A. Jr., and Capelas de Oliveira, E., *The Many Faces of Maxwell, Dirac and Einstein Equations. A Clifford Bundle Approach.* Lecture Notes in Physics **722**, Springer, Heidelberg, 2007.

[3] Hestenes D., and Sobczyk, G., *Clifford Algebras to Geometric Calculus*, D. Reidel Publ. Co., Dordrecht. 1984.

[4 Fernández, V. V., Moya, A. M., and Rodrigues, W. A. Jr., Metric Tensor Vs. Metric Extensors, *Adv. Appl. Clifford Algebras* **11**(S3), 41-48 (2001). [http://www.clifford-algebras.org/]

[5] Moya, A. M, Fernández, V. V., and Rodrigues, W. A. Jr., Metric Clifford Algebra, *Adv. Appl. Clifford Algebras* **11**(S3) 49-68, (2001). [http://www.clifford-algebras.org/]

# 3
# Multiform Functions and Multiform Functionals

ABSTRACT This chapter, of pure mathematical character, deals with the theory of multiform functions which are mappings $\bigwedge V \times ...\bigwedge V \rightarrow \bigwedge V$, and multiform functionals which are mappings $(p,q)$-$extV \rightarrow \bigwedge V$. As a preparation for the general theory, the multiform functions of a real variable are first introduced, defining for those objects the concepts of limit, continuity and derivative. Next, we will introduce multiform functions of multiform variables, defining for them limit, continuity, directional derivatives in the direction of a multiform, derivative mappings in relation to a multiform variable and the concept of $t$-distortions ($t$ being an extensor field) of the derivative mapping, a crucial ingredient for the applications we have in mind. Next, multiform functionals are defined and the key concept of derivatives of induced multiform functionals is given. Finally, we will introduce a derivative operator of induced functionals, called the variational operator, which plays a crucial role in the Lagrangian formulation and in derivation from a variational principle of the motion's equation for the extensor field $\boldsymbol{h}$, representing the distortion field in our theory of the gravitational field. The chapter illustrates the abstract mathematical theory, with several examples worked with details, which hopefully will make the reader confident in the use of this abstract mathematical theory and will permit him to check, with success, the equations of the following chapters.

## 3.1 Multiform Functions of Real Variable

A mapping

$$X : S \rightarrow \bigwedge V, \ (S \subseteq \mathbb{R}), \tag{3.1}$$

is called a *multiform function of real variable*.

For reasons of simplicity, when the image of $X$ is a scalar, a 1-form, a biform, etc..., $X$ is said to be a scalar function, 1-form function, biform function, etc...

V.V. Fernández and W.A. Rodrigues Jr., *Gravitation as a Plastic Distortion of the Lorentz Vacuum*, Fundamental Theories of Physics 168, DOI 10.1007/978-3-642-13589-7_3,

### 3.1.1 Limit and Continuity

The concepts of limit and continuity may be easily introduced following a path analogous to the one used in the theory of ordinary functions.

Then, a multiform $L \in \bigwedge V$ is said to be the limit of $X(\lambda)$ *for* $\lambda \in S$ approaching $\lambda_0 \in S$ iff for any real number $\varepsilon > 0$ there exists a real number $\delta > 0$ such that[1] $\|X(\lambda) - L\| < \varepsilon$, if $0 < |\lambda - \lambda_0| < \delta$. As usual, such a concept will be denoted by $\lim_{\lambda \to \lambda_0} X(\lambda) = L$.

The theorems for the limits of multiform functions are completely analogous to the ones of the theory of ordinary functions, i.e.,

$$\lim_{\lambda \to \lambda_0} X(\lambda) + Y(\lambda) = \lim_{\lambda \to \lambda_0} X(\lambda) + \lim_{\lambda \to \lambda_0} Y(\lambda). \tag{3.2}$$

$$\lim_{\lambda \to \lambda_0} X(\lambda) * Y(\lambda) = \lim_{\lambda \to \lambda_0} X(\lambda) * \lim_{\lambda \to \lambda_0} Y(\lambda), \tag{3.3}$$

where $*$ denotes any of the multiform products, i.e., $(\wedge)$, $(\cdot)$, $(\lrcorner, \llcorner)$ or (*Clifford product*).

$X(\lambda)$ is said to be *continuous* at $\lambda_0 \in S$ iff $\lim_{\lambda \to \lambda_0} X(\lambda) = X(\lambda_0)$, and sum and product of continuous multiform functions are continuous.

### 3.1.2 Derivative

The notion of derivative of a multiform function of a real variable is also formulated in complete analogy to the one used in the theory of ordinary functions. Then, the derivative of $X(\lambda)$ *at* $\lambda_0$ is defined by

$$X'(\lambda_0) := \lim_{\lambda \to \lambda_0} \frac{X(\lambda) - X(\lambda_0)}{\lambda - \lambda_0}. \tag{3.4}$$

The rules for derivations, as expected, are completely analogous to the ones valid for ordinary functions.

$$(X + Y)' = X' + Y', \tag{3.5}$$

$$(X * Y)' = X' * Y + X * Y' \text{ (Leibniz rule)}, \tag{3.6}$$

$$(X \circ \phi)' = (X' \circ \phi)\phi' \text{ (chain rule)}, \tag{3.7}$$

where $*$, as above means $(\wedge)$, $(\cdot)$, $(\lrcorner, \llcorner)$ or (*Clifford product*) and $\phi$ is a real ordinary function ($\phi'$ is the derivative of $\phi$).

In the space of derivable multiform functions of real variable, we introduce the symbol $\frac{d}{d\lambda}$ (derivative operator), by $\frac{d}{d\lambda} X(\lambda) = X'(\lambda)$.

We can easily generalize all the above rules for multiform functions of several real variables. So, any additional comments are not necessary. Now we will introduce the really important objects needed for the presentation of our gravitational theory.

---

[1]For any $A \in \bigwedge V$, $\|X\| := \sqrt{\langle X \cdot X \rangle_0}$

## 3.2 Multiform Functions of Multiform Variables

A mapping $F : \bigwedge^{\diamond} V \to \bigwedge V$ is called a *multiform function of a multiform variable* and, when $F(X)$ is a scalar, a 1-form, a biform, etc..., then $F$ is said to be a scalar function, *a 1-form function, a biform function, etc...*

### 3.2.1 Limit and Continuity

Let $F : \bigwedge^{\diamond} V \to \bigwedge V$. A *multiform* $M \in \bigwedge V$ is said to be the limit of $F(X)$ *for* $X \in \bigwedge^{\diamond} V$ *approaching* $X_0 \in \bigwedge V$ iff for any real $\varepsilon > 0$, there exists a real $\delta > 0$ such that $\|F(X) - M\| < \varepsilon$, if $0 < \|X - X_0\| < \delta$. This will be denoted by $\lim_{X \to X_0} F(X) = M$.

When we are considering scalar functions of multiform variables $\bigwedge^{\diamond} V \ni X \mapsto \Phi(X) \in \mathbb{R}$, the definition $\lim_{X \to X_0} \Phi(X) = \alpha$ is simply the statement: for any real $\varepsilon > 0$ there exists a real $\delta > 0$ such that $|\Phi(X) - \alpha| < \varepsilon$, if $0 < \|X - X_0\| < \delta$.

The limit theorems are analogous to the ones of the theory of ordinary functions, i.e.,

$$\lim_{X \to X_0} F(X) + G(X) = \lim_{X \to X_0} F(X) + \lim_{X \to X_0} G(X). \tag{3.8}$$

$$\lim_{X \to X_0} F(X) * G(X) = \lim_{X \to X_0} F(X) * \lim_{X \to X_0} G(X), \tag{3.9}$$

where $*$ means, as now usual $(\wedge)$, $(\cdot)$, $(\lrcorner, \llcorner)$ or the (Clifford product).

Let $F : \bigwedge^{\diamond} V \to \bigwedge V$ be a multiform function of multiform variable. We say that $F$ is continuous at $X_0 \in \bigwedge V$ iff $\lim_{X \to X_0} F(X) = F(X_0)$.

The sum $X \mapsto (F + G)(X) = F(X) + G(X)$ and any product of continuous multiform functions of multiform variable $X \mapsto (F * G)(X) = F(X) * G(X)$ are also continuous multiform functions of multiform variable.

### 3.2.2 Differentiability

A multiform function of multiform variable $F : \bigwedge^{\diamond} V \to \bigwedge V$ is said to be differentiable at $X_0 \in \bigwedge V$ iff it exists $f_{X_0} \in extV$, such that

$$\lim_{X \to X_0} \frac{F(X) - F(X_0) - f_{X_0}(X - X_0)}{\|X - X_0\|} = 0. \tag{3.10}$$

If such $f_{X_0}$ *exists,* it must be unique and it will be called the *differential* of $F$ *at* $X_0$.

### 3.2.3 The Directional Derivative $A \cdot \partial_X$

Let $F : \bigwedge^{\diamond} V \to \bigwedge V$ be differentiable at $X$. Take a real number $\lambda \neq 0$ and an arbitrary multiform $A$. Denote also by $\langle \rangle_X : \bigwedge V \to \bigwedge^{\diamond} V$. Then, the limit

$$\lim_{\lambda \to 0} \frac{F(X + \lambda \langle A \rangle_X) - F(X)}{\lambda} \tag{3.11}$$

exists. It will be denoted by $F'_A(X)$ (or, $A \cdot \partial_X F(X)$)[2] and called the directional derivative of $F$ *at $X$ in the direction of the multiform $A$*. The operator $A \cdot \partial_X$ is said to be the directional derivative operator in the direction of $A$. We have:

$$F'_A(X) = A \cdot \partial_X F(X) := \lim_{\lambda \to 0} \frac{F(X + \lambda \langle A \rangle_X) - F(X)}{\lambda}, \tag{3.12}$$

or in an equivalent form

$$F'_A(X) = A \cdot \partial_X F(X) = \frac{d}{d\lambda} F(X + \lambda \langle A \rangle_X) \bigg|_{\lambda=0}. \tag{2.10}$$

The directional derivative of a differentiable multiform function $F : \bigwedge^\diamond V \to \bigwedge V$ is linear with respect to the direction multiform, i.e., for all $\alpha, \beta \in \mathbb{R}$ and $A, B \in \bigwedge V$

$$F'_{\alpha A + \beta B}(X) = \alpha F'_A(X) + \beta F'_B(X), \tag{3.13}$$

or yet

$$(\alpha A + \beta B) \cdot \partial_X F(X) = \alpha A \cdot \partial_X F(X) + \beta A \cdot \partial_X F(X). \tag{3.14}$$

We will give now the main properties of the directional derivative.

Let $F, G : \bigwedge^\diamond V \to \bigwedge V$ be differentiable, the sum $X \mapsto (F + G)(X) = F(X) + G(X)$ and the products $X \mapsto (F * G)(X) = F(X) * G(X)$, where $*$ means $(\wedge)$, $(\cdot)$, $(\lrcorner, \llcorner)$ or (*Clifford product*), are also differentiable and we have

$$(F + G)'_A(X) = F'_A(X) + G'_A(X), \tag{3.15}$$

$$(F * G)'_A(X) = F'_A(X) * G(X) + F(X) * G'_A(X) \text{ (Leibniz rule)}. \tag{3.16}$$

Chain Rules

Let $F, G : \bigwedge^\diamond V \to \bigwedge V$ be differentiable. The composition $X \mapsto (F \circ G)(X) = F(\langle G(X) \rangle_Y)$, with $Y \in \mathrm{dom} F$ is differentiable and

$$(F \circ G)'_A(X) = F'_{G'_A(X)}(\langle G(X) \rangle_Y). \tag{3.17}$$

Let $X \mapsto F(X)$ and $\lambda \mapsto X(\lambda)$ be differentiable multiform functions, the first one of multiform variable and the second of real variable. The composition $\lambda \mapsto (F \circ X)(\lambda) = F(\langle X(\lambda) \rangle_Y)$, with $Y \in \mathrm{dom} F$ is a differentiable multiform function of a real variable and

$$(F \circ X)'(\lambda) = F'_{X'(\lambda)}(\langle X(\lambda) \rangle_Y). \tag{3.18}$$

Let $\phi : \mathbb{R} \to \mathbb{R}$ and $\Psi : \bigwedge^\diamond V \to \mathbb{R}$, respectively an ordinary differentiable function and a scalar differentiable multiform function of multiform variables. The composition $\phi \circ \Psi : \bigwedge^\diamond V \to \mathbb{R}$ such that $(\phi \circ \Psi)(X) = \phi(\Psi(X))$ is differentiable multiform function of multiform variable and

$$(\phi \circ \Psi)'_A(X) = \phi'(\Psi(X)) \Psi'_A(X). \tag{3.19}$$

[2]For some few special cases we use some special symbols.

In resume, we have:

$$A \cdot \partial_X (F + G)(X) = A \cdot \partial_X F(X) + A \cdot \partial_X G(X). \tag{3.20}$$

$$A \cdot \partial_X (F * G)(X) = A \cdot \partial_X F(X) * G(X) + F(X) * A \cdot \partial_X G(X). \tag{3.21}$$

$$A \cdot \partial_X (F \circ G)(X) = A \cdot \partial_X G(X) \cdot \partial_Y F(\langle G(X)\rangle_Y), \text{ with } Y \in \text{dom}F. \tag{3.22}$$

$$\frac{d}{d\lambda}(F \circ X)(\lambda) = \frac{d}{d\lambda} X(\lambda) \cdot \partial_Y F(\langle X(\lambda)\rangle_Y), \text{ with } Y \in \text{dom}F. \tag{3.23}$$

$$A \cdot \partial_X (\phi \circ \Psi)(X) = \frac{d}{d\mu}\phi(\Psi(X)) A \cdot \partial_X \Psi(X). \tag{3.24}$$

### *3.2.4 The Derivative Mapping* $\partial_X$

Let $F : \bigwedge^\diamond V \to \bigwedge V$ be differentiable at $X$. Take any pair of arbitrary reciprocal bases for $V$, say $(\{\varepsilon^j\}, \{\varepsilon_j\})$, $(\varepsilon^j \cdot \varepsilon_i = \delta_i^j)$. Then, it exists a well defined multiform function (i.e., it is independent of the pair $(\{\varepsilon^j\}, \{\varepsilon_j\})$ depending only on $F$) called the derivative of $F$ *at* $X$, and given by

$$F'(X) = \partial_X F(X) := \sum_J \frac{1}{\nu(J)!} \varepsilon^J F'_{\varepsilon_J}(X) = \sum_J \frac{1}{\nu(J)!} \varepsilon_J F'_{\varepsilon^J}(X). \tag{3.25}$$

The main properties of $\partial_X F$ are:

(i) Let $X \mapsto \Phi(X)$ be a scalar multiform function, then

$$A \cdot \partial_X \Phi(X) = A \cdot (\partial_X \Phi(X)). \tag{3.26}$$

(ii) Let $X \mapsto F(X)$, $X \mapsto G(X)$ and $X \mapsto \Phi(X)$ be differentiable multiform functions. Then,

$$\partial_X (F + G)(X) = \partial_X F(X) + \partial_X G(X). \tag{3.27}$$

$$\partial_X (\Phi G)(X) = \partial_X \Phi(X) G(X) + \Phi(X) \partial_X G(X) \text{ (Leibniz rule)}. \tag{3.28}$$

### *3.2.5 Examples*

Now we will present some examples which illustrate some of the most important formulas of the calculus of multiform functions of multiform variables and which will be used in the following sections.

**Example 3.1** Let $\bigwedge^\diamond V \ni X \mapsto X \cdot X \in \mathbb{R}$. Let us calculate $A \cdot \partial_X (X \cdot X)$ and $\partial_X (X \cdot X)$.

$$\begin{aligned} A \cdot \partial_X (X \cdot X) &= \frac{d}{d\lambda}(X + \lambda \langle A\rangle_X) \cdot (X + \lambda \langle A\rangle_X)\Big|_{\lambda=0} \\ &= \frac{d}{d\lambda}(X \cdot X + 2\lambda \langle A\rangle_X \cdot X + \lambda^2 \langle A\rangle_X \cdot \langle A\rangle_X)\Big|_{\lambda=0}, \end{aligned}$$

$$A \cdot \partial_X (X \cdot X) = 2 \langle A\rangle_X \cdot X = 2A \cdot X. \tag{3.29a}$$

$$\partial_X (X \cdot X) = \sum_J \frac{1}{\nu(J)!} \varepsilon^J \varepsilon_J \cdot \partial_X (X \cdot X) = \sum_J \frac{1}{\nu(J)!} \varepsilon^J 2(\varepsilon_J \cdot X) = 2X. \tag{3.29b}$$

**Example 3.2** Let $\bigwedge^{\diamond} V \ni X \mapsto B \cdot X \in \mathbb{R}$, with $B \in \bigwedge V$. Let us calculate $A \cdot \partial_X (B \cdot X)$ and $\partial_X (B \cdot X)$.

$$A \cdot \partial_X (B \cdot X) = \frac{d}{d\lambda} B \cdot (X + \lambda \langle A \rangle_X) \bigg|_{\lambda=0} = B \cdot \langle A \rangle_X = A \cdot \langle B \rangle_X \,, \tag{3.30a}$$

$$\partial_X (B \cdot X) = \sum_J \frac{1}{\nu(J)!} \varepsilon^J \varepsilon_J \cdot \partial_X (B \cdot X)$$
$$= \sum_J \frac{1}{\nu(J)!} \varepsilon^J (\varepsilon_J \cdot \langle B \rangle_X) = \langle B \rangle_X \,. \tag{3.30b}$$

**Example 3.3** Let $\bigwedge^{\diamond} V \ni X \mapsto (BXC) \cdot X \in \mathbb{R}$, with $B, C \in \bigwedge V$. Let us calculate: $A \cdot \partial_X ((BXC) \cdot X)$ and $\partial_X ((BXC) \cdot X)$.

$$A \cdot \partial_X ((BXC) \cdot X) = \frac{d}{d\lambda} (B(X + \lambda \langle A \rangle_X) C) \cdot (X + \lambda \langle A \rangle_X) \bigg|_{\lambda=0}$$
$$= \frac{d}{d\lambda} ((BXC) \cdot X + \lambda (BXC \cdot \langle A \rangle_X)) \bigg|_{\lambda=0}$$
$$+ \frac{d}{d\lambda} (\lambda (B \langle A \rangle_X C) \cdot X) + \lambda^2 (B \langle A \rangle_X C) \cdot \langle A \rangle_X)) \bigg|_{\lambda=0} \tag{3.31a}$$
$$= (BXC) \cdot \langle A \rangle_X + (B \langle A \rangle_X C) \cdot X \tag{3.31b}$$
$$= \langle A \rangle_X \cdot (BXC + \widetilde{B} X \widetilde{C}),$$
$$A \cdot \partial_X ((BXC) \cdot X) = A \cdot \left\langle BXC + \widetilde{B} X \widetilde{C} \right\rangle_X .$$
$$\partial_X ((BXC) \cdot X) = \sum_J \frac{1}{\nu(J)!} \varepsilon^J \varepsilon_J \cdot \partial_X (BXC \cdot X)$$
$$= \sum_J \frac{1}{\nu(J)!} \varepsilon^J (\varepsilon_J \cdot \left\langle BXC + \widetilde{B} X \widetilde{C} \right\rangle_X),$$
$$\partial_X ((BXC) \cdot X) = \left\langle BXC + \widetilde{B} X \widetilde{C} \right\rangle_X . \tag{3.31c}$$

In this example, we essentially utilized the nontrivial multiform identity $(AXB) \cdot Y = X \cdot (\widetilde{A} Y \widetilde{B})$.

**Example 3.4** Consider $x \wedge B$, with $x \in \bigwedge^1 V$ and $B \in \bigwedge^2 V$. Then, $x \wedge B \in \bigwedge^3 V$. Let us calculate the directional derivative $a \cdot \partial_x (x \wedge B)$, the *divergent* $\partial_x \lrcorner (x \wedge B)$, the *rotational* $\partial_x \wedge (x \wedge B)$ and the gradient $\partial_x(x \wedge B)$.

$$a \cdot \partial_x(x \wedge B) = \frac{d}{d\lambda}(x + \lambda a) \wedge B\Big|_{\lambda=0} = a \wedge B, \tag{3.32a}$$

$$\partial_x \lrcorner (x \wedge B) = \sum_{j=1}^{n} \varepsilon^j \lrcorner \varepsilon_j \cdot \partial_x(x \wedge B) = \sum_{j=1}^{n} \varepsilon^j \lrcorner (\varepsilon_j \wedge B) = (n-2)B. \tag{3.32b}$$

$$\partial_x \wedge (x \wedge B) = \sum_{j=1}^{n} \varepsilon^j \wedge \varepsilon_j \cdot \partial_x(x \wedge B) = \sum_{j=1}^{n} \varepsilon^j \wedge (\varepsilon_j \wedge B) = 0. \tag{3.32c}$$

$$\begin{aligned} \partial_x(x \wedge B) &= \sum_{j=1}^{n} \varepsilon^j \varepsilon_j \cdot \partial_x(x \wedge B) = \sum_{j=1}^{n} \varepsilon^j (\varepsilon_j \wedge B) \\ &= \sum_{j=1}^{n} (\varepsilon^j \lrcorner (\varepsilon_j \wedge B) + \varepsilon^j \wedge (\varepsilon_j \wedge B)), \end{aligned} \tag{3.32d}$$

$$\partial_x(x \wedge B) = (n-2)B, \qquad (n = \dim V). \tag{3.32e}$$

**Example 3.5** Consider $x \lrcorner\, B \in \bigwedge^2 V$. Let us calculate $a \cdot \partial_x(x \lrcorner B)$, $\partial_x \lrcorner (x \lrcorner B)$, $\partial_x \wedge (x \lrcorner B)$ and $\partial_x(x \lrcorner B)$

$$a \cdot \partial_x(x \lrcorner B) = \frac{d}{d\lambda}(x + \lambda a) \lrcorner B\Big|_{\lambda=0} = a \lrcorner B. \tag{3.33a}$$

$$\partial_x \lrcorner (x \lrcorner B) = \sum_{j=1}^{n} \varepsilon^j \lrcorner \varepsilon_j \cdot \partial_x(x \lrcorner B) = \sum_{j=1}^{n} \varepsilon^j \lrcorner (\varepsilon_j \lrcorner B) = \sum_{j=1}^{n} (\varepsilon^j \wedge \varepsilon_j) \lrcorner\, B = 0, \tag{3.33b}$$

$$\partial_x \wedge (x \lrcorner B) = \sum_{j=1}^{n} \varepsilon^j \wedge \varepsilon_j \cdot \partial_x(x \lrcorner B) = \sum_{j=1}^{n} \varepsilon^j \wedge (\varepsilon_j \lrcorner B) = 2B. \tag{3.33c}$$

$$\partial_x(x \lrcorner B) = \sum_{j=1}^{n} \varepsilon^j \varepsilon_j \cdot \partial_x(x \lrcorner B) = \sum_{j=1}^{n} \varepsilon^j (\varepsilon_j \lrcorner B) \tag{3.33d}$$

$$\partial_x(x \lrcorner B) = \sum_{j=1}^{n} (\varepsilon^j \lrcorner (\varepsilon_j \lrcorner B) + \varepsilon^j \wedge (\varepsilon_j \lrcorner B)) = 2B. \qquad (n = \dim V) \tag{3.33e}$$

**Example 3.6 Let** $X \in \bigwedge^r V$ and consider the multiform function $F : \bigwedge^r \ni X \mapsto X \wedge \star X \in \bigwedge^n V$. Let us calculate the directional derivative $W \cdot \partial_X F$, with $W \in \bigwedge^r V$ and $\partial_X F$. We have

$$\begin{aligned} W \cdot \partial_X F(X) &= \lim_{\lambda \to 0} \frac{(X + \lambda W) \wedge \star(X + \lambda W) - X \wedge \star X}{\lambda} \\ &= 2W \wedge \star X. \end{aligned} \tag{3.34}$$

On the other hand, using Eq.(3.25), we have:

$$\begin{aligned} \partial_X F(X) &= \sum (\frac{1}{r!} \varepsilon^{j_1} \wedge ... \wedge \varepsilon^{j_r})[(\varepsilon_{j_1} \wedge ... \wedge \varepsilon_{j_r}) \cdot \partial_X F(X)] \\ &= 2 \sum \frac{1}{r!} \varepsilon^{j_1} \wedge ... \wedge \varepsilon^{j_r} [(\varepsilon_{j_1} \wedge ... \wedge \varepsilon_{j_r}) \wedge \star X] \\ &= 2 \sum \frac{1}{r!} \varepsilon^{j_1} \wedge ... \wedge \varepsilon^{j_r} (-1)^{r(r-1)} [\star((\widetilde{\varepsilon_{j_1} \wedge ... \wedge \varepsilon_{j_r}}) \lrcorner X))] \\ &= 2 \sum \frac{1}{r!} \varepsilon^{j_1} \wedge ... \wedge \varepsilon^{j_r} [(\varepsilon_{j_1} \wedge ... \wedge \varepsilon_{j_r}) \cdot X) \tau] \\ &= 2X\tau = 2(-1)^{\frac{r}{2}(r-1)} \tilde{X} \lrcorner \tau = 2(-1)^{\frac{r}{2}(r-1)} \star X. \end{aligned}$$

**Remark 3.1** Sometimes, the directional derivative of a multiform function $F : \bigwedge^r \ni X \mapsto X \wedge \star X \in \bigwedge^n$ in the direction of $W = \boldsymbol{\delta} X \in \bigwedge^r V$ is written as

$$\boldsymbol{\delta} F := \boldsymbol{\delta} X \wedge \frac{\partial F}{\partial X} \tag{3.35}$$

and called the *variational derivative* of $F$ and $\frac{\partial F}{\partial X}$ is called the algebraic derivative of $F$. Now, since $\boldsymbol{\delta} F = W \cdot \partial_X$ we see that we have the identification for our multiform function

$$\boldsymbol{\delta} X \cdot \partial_X F = (-1)^{\frac{r}{2}(r-1)} \boldsymbol{\delta} X \wedge \frac{\partial F}{\partial X}. \tag{3.36}$$

### *3.2.6 The Operators $\partial_X *$ and their t-distortions*

We will introduce now the linear differential operators $\partial_X *$ defined by

$$\partial_X * F(X) := \sum_J \frac{1}{\nu(J)!} \varepsilon^J * \varepsilon_J \cdot \partial_X F(X), \tag{3.37}$$

where $*$ denotes any one of the multiform products $(\wedge)$, $(\cdot)$, $(\lrcorner, \llcorner)$ or (*Clifford product*).

Those linear differentiable operators are well defined since the multiforms in the second member of Eq.(3.37) depends only on the multiform function $F$ and do not depend on the pair of arbitrary reciprocal basis used.

Note that, if $*$ refers to the Clifford product then the operator $\partial_X *$ is precisely the operator $\partial_X$ (i.e., the derivative operator). Sometimes, $\partial_X$ is called the *gradient operator* and $\partial_X F$ is said to be the *gradient of F.*

Also, the operator $\partial_X \wedge$ is called the *rotational operator* and $\partial_X \wedge F$ is said to be the *rotational of F.*

The operators $\partial_X \cdot$ and $\partial_X \lrcorner$ are called divergent operators, $\partial_X \cdot F$ is said to be the *scalar divergent and* $\partial_X \lrcorner F$ is said to be the *contracted divergent.*

Let $t \in (1,1)$-*extV*. We define the linear differential operators $\underline{t}(\partial_X)*$ by

$$\underline{t}(\partial_X) * F(X) := \sum_J \frac{1}{\nu(J)!} \underline{t}(\varepsilon^J) * \varepsilon_J \cdot \partial_X F(X), \tag{3.38}$$

where $*$ is any one of the multiform products $(\wedge)$, $(\cdot)$, $(\lrcorner, \llcorner)$ or (*Clifford product*).

Those linear differential operators are well defined since they depend only on $F$ and do not depend on the pair of reciprocal basis $(\{\varepsilon^j\}, \{\varepsilon_j\})$. We call $\underline{t}(\partial_X)*$ a $t$-distortion of $\partial_X *$. Special cases have special names, i.e., we call $\underline{t}(\partial_X) \wedge F$ the $t$-rotational of $F$, $\underline{t}(\partial_X) \cdot F$ is the $t$-scalar divergent of $F$ and $\underline{t}(\partial_X) \lrcorner F$ is *t-contracted divergent of F*. Finally, $\underline{t}(\partial_X)$ is the *t- gradient operator and* $\underline{t}(\partial_X)F$ is read as the *t-gradient of F*.

## 3.3 Multiform Functionals $\mathcal{F}_{(X^1,\dots,X^k)}[t]$

Let

$$F : \underbrace{\bigwedge^q V \times \cdots \times \bigwedge^q V}_{k \text{ copies}} \to \bigwedge V,$$
$$(X^1, \dots, X^k) \mapsto F(X^1, \dots, X^k) = F_{(X^1,\dots,X^k)}. \tag{3.39}$$

A mapping

$$\mathcal{F}_{(X^1,\dots,X^k)} : (p,q)\text{-}extV \to \bigwedge V,$$
$$t \mapsto \mathcal{F}_{(X^1,\dots,X^k)}[t] := F[t(X^1), \dots, t(X^k)], \tag{3.40}$$

will be called a *multiform functional $(p,q)$-extensor variable induced by the function $F$*. We say also that $F$ is the *generator of $\mathcal{F}_{(X^1,\dots,X^k)}$*.

When $\mathcal{F}[t]$ is scalar valued, 1-form valued, biform valued, etc..., $\mathcal{F}$ is said to be a *scalar functional, 1-form functional, biform functional, etc..., of a $(p,q)$-extensor variable.*

### 3.3.1 Derivatives of Induced Multiform Functionals

Let $(X^1, \dots, X^k) \mapsto F(X^1, \dots, X^k)$ be a differentiable multiform function, i.e., differentiable with respect to each one of its $q$-multiform variables $(X^1, \dots, X^k)$. Then, the partial derivatives of $F$, $\partial_{X^1} F, \dots, \partial_{X^k} F$ exist. We can then construct exactly $k$ multiform functionals of a $(p,q)$-extensor variable $t \in (p,q)\text{-}extV$ which are induced by them, namely

$$t \mapsto (\partial_{X^1} F)_{(X^1,\dots,X^k)}[t] := \partial_{t(X^1)} F[t(X^1), \dots, t(X^k)], \dots,$$
$$t \mapsto (\partial_{X^k} F)_{(X^1,\dots,X^k)}[t] := \partial_{t(X^k)} F[t(X^1), \dots, t(X^k)]. \tag{3.41}$$

#### The $A$-Directional $A \cdot \partial_t$ Derivative of a Multiform Functional

Let $A \in \bigwedge V$ be an arbitrary multiform. The multiform functional of $(p,q)$-extensor variable $t$,

$$(p,q)\text{-}extV \ni t \mapsto A \cdot \partial_t \mathcal{F}_{(X^1,\dots,X^k)}[t] \in \bigwedge V \tag{3.42}$$

such

$$\begin{aligned} A \cdot \partial_t \mathcal{F}_{(X^1,\dots,X^k)}[t] &:= A \cdot X^1 (\partial_{X^1} F)_{(X^1,\dots,X^k)}[t] + \cdots \\ &\quad + A \cdot X^k (\partial_{X^k} F)_{(X^1,\dots,X^k)}[t] \\ &= \sum_{i=1}^k A \cdot X^i (\partial_{X^i} F)_{(X^1,\dots,X^k)}[t] \\ &= \sum_{i=1}^k A \cdot X^i \partial_{t(X^i)} F[t(X^1), \dots, t(X^k)] \end{aligned} \tag{3.43}$$

is said to be [3] the *A-directional derivative of the functional* $\mathcal{F}_{(X^1,\ldots,X^k)}[t]$. A convenient notation, in order to write short formulas, is $\mathcal{F}'_{(X^1,\ldots,X^k)A}[t]$, i.e., we have

$$\mathcal{F}'_{(X^1,\ldots,X^k)A}[t] \equiv A\cdot\partial_t\mathcal{F}_{(X^1,\ldots,X^k)}[t]. \tag{3.44}$$

Observe that $A\cdot\partial_t$ is linear with respect to the direction multiform $A$, i.e., for all $\alpha,\beta\in\mathbb{R}$ and $A,B\in\bigwedge V$,

$$(\alpha A+\beta B)\cdot\partial_t = \alpha A\cdot\partial_t + \beta B\cdot\partial_t. \tag{3.45}$$

The operator $A\cdot\partial_t$ possesses three important properties:

$$\begin{aligned}
A\cdot\partial_t\ (\mathcal{F}_{(X^1,\ldots,X^k)}+\mathcal{G}_{(X^1,\ldots,X^k)})[t] &= A\cdot\partial_t\ \mathcal{F}_{(X^1,\ldots,X^k)}[t] && (3.46)\\
&\quad + A\cdot\partial_t\ \mathcal{G}_{(X^1,\ldots,X^k)}[t].\\
A\cdot\partial_t\ (\alpha\mathcal{F}_{(X^1,\ldots,X^k)})[t] &= \alpha(A\cdot\partial_t\ \mathcal{F}_{(X^1,\ldots,X^k)}[t]) && (3.47)\\
A\cdot\partial_t(\mathcal{F}_{(X^1,\ldots,X^k)}B)[t] &= (A\cdot\partial_t\ \mathcal{F}_{(X^1,\ldots,X^k)}[t])B, && (3.48)
\end{aligned}$$

where $\mathcal{F}_{(X^1,\ldots,X^k)}$ and $\mathcal{G}_{(X^1,\ldots,X^k)}$ are the multiform functionals induced by $F$ and $G$, and $\alpha\in\mathbb{R}$ and $B\in\bigwedge V$.

### The Operators $\partial_t *$

Let $(\{\varepsilon^j\},\{\varepsilon_j\})$ be a pair of arbitrary reciprocal bases for $V$ and let $*$ be any of the multiform products as before. We define

$$\partial_t * \mathcal{F}_{(X^1,\ldots,X^k)}[t] := \frac{1}{p!}\varepsilon^{j_1}\wedge\ldots\wedge\varepsilon^{j_p} * \mathcal{F}'_{(X^1,\ldots,X^k)\varepsilon_{j_1}\wedge\ldots\wedge\varepsilon_{j_p}}[t]. \tag{3.49}$$

Observe first that $\partial_t * \mathcal{F}_{(A^1,\ldots,A^k)}[t]$ does not depend on the pair of reciprocal bases and that taking into account Eq.(3.43) and Eq.(3.41), we have

$$\begin{aligned}
&\partial_t * \mathcal{F}_{(X^1,\ldots,X^k)}[t]\\
&= \frac{1}{p!}\varepsilon^{j_1}\wedge\ldots\wedge\varepsilon^{j_p} * \sum_{i=1}^{k}(\varepsilon_{j_1}\wedge\ldots\wedge\varepsilon_{j_p})\cdot X^i(\partial_{X^i}F)_{(X^1,\ldots,X^k)}[t]\\
&= \sum_{i=1}^{k}\frac{1}{p!}(\varepsilon_{j_1}\wedge\ldots\wedge\varepsilon_{j_p})\cdot X^i(\varepsilon^{j_1}\wedge\ldots\wedge\varepsilon^{j_p}) * (\partial_{X^i}F)_{(A^1,\ldots,A^k)}[t]\\
&= \sum_{i=1}^{k}X^i * (\partial_{X^i}F)_{(X^1,\ldots,X^k)}[t]\\
&= \sum_{i=1}^{k}X^i * \partial_{t(X^k)}F[t(X^1),\ldots,t(X^k)].
\end{aligned} \tag{3.50}$$

We call the objects $\partial_t\wedge\mathcal{F}_{(A^1,\ldots,A^k)}[t]$, $\partial_t\cdot\mathcal{F}_{(A^1,\ldots,A^k)}[t]$, $\partial_t\lrcorner\mathcal{F}_{(A^1,\ldots,A^k)}[t]$ and $\partial_t\mathcal{F}_{(A^1,\ldots,A^k)}[t]$ respectively the *rotational*, the *divergence*, the l*eft contracted divergence and the gradient* (or simply the derivative) of $\mathcal{F}_{(X^1,\ldots,X^k)}$ with respect to $t$ [1].

---

[3]Note that in [1] we used the notation $\partial_{t(A)}$ in place of $A\cdot\partial_t$ and called that operator the directional derivative of the functional $\mathcal{F}_{(X^1,\ldots,X^k)}[t]$ in the direction of $A$. We think now that the new denomination is more appropriate, something that will become clearer when we calculate some examples.

Examples

We will present now some illustrative examples of derivatives of multiform functionals that will be needed in Chapter 5 (and Appendix C), where we shall derive the equations of motion of the gravitational field from the variational principle.

**Example 3.7** Let $\boldsymbol{h} \in (1,1)\text{-}extV$ and consider the scalar functional $\boldsymbol{h} \mapsto \boldsymbol{h}(b) \cdot \boldsymbol{h}(c)$ and the biform functional $\boldsymbol{l} \mapsto \boldsymbol{l}(b) \wedge \boldsymbol{l}(c)$, with $b, c \in \bigwedge^1 V$. The first has as generator function, the scalar function of two 1-form variables $(x, y) \mapsto x \cdot y$ and the generator of the second functional is the biform function of two 1-form variables $(x, y) \mapsto x \wedge y$. Let us calculate the derivative of those functionals, with respect to $\boldsymbol{h}$ in the direction of $a \in \bigwedge^1 V$.

$$\begin{aligned} a \cdot \partial_h(\boldsymbol{h}(b) \cdot \boldsymbol{h}(c)) &= a \cdot b\partial_{h(b)}(\boldsymbol{h}(b) \cdot \boldsymbol{h}(c)) + a \cdot c\partial_{h(c)}(\boldsymbol{h}(b) \cdot \boldsymbol{h}(c)) \\ &= a \cdot b\boldsymbol{h}(c) + a \cdot c\boldsymbol{h}(b), \end{aligned} \tag{3.51}$$

where we utilized $\partial_x(x \cdot y) = y$.

Also

$$\begin{aligned} a \cdot \partial_h(\boldsymbol{h}(b) \wedge \boldsymbol{h}(c)) &= a \cdot b\partial_{h(b)}(\boldsymbol{h}(b) \wedge \boldsymbol{h}(c)) + a \cdot c\partial_{h(c)}(\boldsymbol{h}(b) \wedge \boldsymbol{h}(c)) \\ &= a \cdot b(n-1)\boldsymbol{h}(c) - a \cdot c(n-1)\boldsymbol{h}(b) \\ &= (n-1)(a \cdot b\boldsymbol{h}(c) - a \cdot c\boldsymbol{h}(b)) \\ &= (n-1)\boldsymbol{h}(a \cdot bc - a \cdot cb) \\ &= (n-1)\boldsymbol{h}(a \lrcorner (b \wedge c)), \end{aligned} \tag{3.52}$$

where we used $\partial_x(x \wedge y) = (n-1)y$.

Eq.(3.52) has a very useful generalization. The derivative of the $k$-form functional $\boldsymbol{h} \mapsto \underline{\boldsymbol{h}}(a^1 \wedge \ldots \wedge a^k) = \boldsymbol{h}(a^1) \wedge \ldots \wedge \boldsymbol{h}(a^k)$, with $a^1, \ldots, a^k \in \bigwedge^1 V$ is

$$a \cdot \partial_h \underline{\boldsymbol{h}}(a^1 \wedge \ldots \wedge a^k) = (n-k+1)\underline{\boldsymbol{h}}(a \lrcorner (a^1 \wedge \ldots \wedge a^k)). \tag{3.53}$$

**Example 3.8** Let $t \in (1,1)\text{-}extV$. The trace of $t$ (i.e., $t \mapsto \mathrm{tr}[t] = t(\varepsilon^j) \cdot \varepsilon_j$) is a scalar functional of $t$, whose generator is the scalar function of $n$ 1-form variables, $(x^1, \ldots, x^n) \mapsto x^1 \cdot \varepsilon_1 + \cdots + x^n \cdot \varepsilon_n$, and $\mathrm{bif}[t] = t(\varepsilon^j \wedge \varepsilon_j)$ is the biform functional of $t$, whose generator is the biform function of $n$ 1-form variables, $(x^1, \ldots, x^n) \mapsto x^1 \wedge \varepsilon_1 + \cdots + x^n \wedge \varepsilon_n$. Let us calculate $a \cdot \partial_t \mathrm{tr}[t]$ and $\partial_{\boldsymbol{h}(a)} \mathrm{bif}[t]$. We have

$$\begin{aligned} a \cdot \partial_h \mathrm{tr}[t] &= a \cdot \varepsilon^1 \partial_{t(\varepsilon^1)}(t(\varepsilon^1) \cdot \varepsilon_1 + \cdots + t(\varepsilon^n) \cdot \varepsilon_n) + \cdots \\ &\quad + a \cdot \varepsilon^n \partial_{t(\varepsilon^n)}(t(\varepsilon^1) \cdot \varepsilon_1 + \cdots + t(\varepsilon^n) \cdot \varepsilon_n) \\ &= a \cdot \varepsilon^1 \varepsilon_1 + \cdots + a \cdot \varepsilon^n \varepsilon_n = a, \end{aligned} \tag{3.54}$$

and

$$\begin{aligned} a \cdot \partial_h \mathrm{bif}[t] &= a \cdot \varepsilon^1 \partial_{t(\varepsilon^1)}(t(\varepsilon^1) \wedge \varepsilon_1 + \cdots + t(\varepsilon^n) \wedge \varepsilon_n) + \cdots \\ &\quad + a \cdot \varepsilon^n \partial_{t(\varepsilon^n)}(t(\varepsilon^1) \wedge \varepsilon_1 + \cdots + t(\varepsilon^n) \wedge \varepsilon_n) \\ &= a \cdot \varepsilon^1 (n-1)\varepsilon_1 + \cdots + a \cdot \varepsilon^n (n-1)\varepsilon_n = (n-1)a. \end{aligned} \tag{3.55}$$

**Example 3.9** Let $\boldsymbol{h} \in (1,1)$-$extV$, and consider the 1-form functional $\boldsymbol{h} \mapsto \boldsymbol{h}^\dagger(b)$, with $b \in \bigwedge^1 V$, and the pseudo-scalar functional $\boldsymbol{h} \mapsto \underline{\boldsymbol{h}}(\tau)$, with $\tau \in \bigwedge^n V$. Let us calculate the derivatives of those functionals.

$$\begin{aligned} a \cdot \partial_{\boldsymbol{h}} \boldsymbol{h}^\dagger(b) &= a \cdot \partial_{\boldsymbol{h}}(\boldsymbol{h}^\dagger(b) \cdot \varepsilon^k \varepsilon_k) = a \cdot \partial_{\boldsymbol{h}}(b \cdot \boldsymbol{h}(\varepsilon^k)\varepsilon_k) \\ &= a \cdot \varepsilon^1 \partial_{\boldsymbol{h}(\varepsilon^1)}(b \cdot \boldsymbol{h}(\varepsilon^k)\varepsilon_k) + \cdots + a \cdot \varepsilon^n \partial_{\boldsymbol{h}(\varepsilon^n)}(b \cdot \boldsymbol{h}(\varepsilon^k)\varepsilon_k) \\ &= a \cdot \varepsilon^1 (b\varepsilon_1) + \cdots + a \cdot \varepsilon^n (b\varepsilon_n) \\ &= b(a \cdot \varepsilon^1 \varepsilon_1 + \cdots + a \cdot \varepsilon^n \varepsilon_n) = ba, \end{aligned} \tag{3.56}$$

where we used a formula for the expansion of 1-forms, in order to get $\boldsymbol{h}^\dagger(b)$ as a 1-form functional of $\boldsymbol{h}$, the scalar product condition involving $\boldsymbol{h}$ and $\boldsymbol{h}^\dagger$ (i.e., $\boldsymbol{h}^\dagger(x) \cdot y = x \cdot \boldsymbol{h}(y)$) and the formula $\partial_x(b \cdot x)c = bc$.

Also,

$$\begin{aligned} a \cdot \partial_{\boldsymbol{h}} \underline{\boldsymbol{h}}(\tau) &= a \cdot \partial_{\boldsymbol{h}} \underline{\boldsymbol{h}}((\tau \cdot \varepsilon_1 \wedge \ldots \wedge \varepsilon_n)\varepsilon^1 \wedge \ldots \wedge \varepsilon^n) \\ &= (\tau \cdot \varepsilon_1 \wedge \ldots \wedge \varepsilon_n) a \cdot \partial_{\boldsymbol{h}} \underline{\boldsymbol{h}}(\varepsilon^1 \wedge \ldots \wedge \varepsilon^n) \\ &= (\tau \cdot \varepsilon_1 \wedge \ldots \wedge \varepsilon_n) \underline{\boldsymbol{h}}(a \lrcorner (\varepsilon^1 \wedge \ldots \wedge \varepsilon^n)) \\ &= \underline{\boldsymbol{h}}(a \lrcorner (\tau \cdot \varepsilon_1 \wedge \ldots \wedge \varepsilon_n)\varepsilon^1 \wedge \ldots \wedge \varepsilon^n) \\ &= \underline{\boldsymbol{h}}(a \lrcorner \tau) = \underline{\boldsymbol{h}}(a\tau), \end{aligned} \tag{3.57}$$

where we used the formula for expansion of pseudo-scalars, Eq.(3.47) and Eq.(3.53).

**Example 3.10** Let $\boldsymbol{h} \in (1,1)$-$extV$, the determinant of $\boldsymbol{h}$ is a well defined scalar functional, $\boldsymbol{h} \mapsto \det[\boldsymbol{h}] = \underline{\boldsymbol{h}}(\varepsilon^1 \wedge \ldots \wedge \varepsilon^n) \cdot (\varepsilon_1 \wedge \ldots \wedge \varepsilon_n)$. Let us calculate $\partial_{\boldsymbol{h}(a)} \det[\boldsymbol{h}]$.

$$\begin{aligned} a \cdot \partial_{\boldsymbol{h}} \det[\boldsymbol{h}] &= a \cdot \partial_{\boldsymbol{h}}(\underline{\boldsymbol{h}}(\tau)\tau) = (a \cdot \partial_{\boldsymbol{h}} \underline{\boldsymbol{h}}(\tau))\tau^{-1} \\ &= \underline{\boldsymbol{h}}(a\tau)\tau^{-1} = \det[\boldsymbol{h}]\boldsymbol{h}^{\clubsuit}(a), \end{aligned} \tag{3.58}$$

where we used Eq.(3.48), Eq.(3.57) and the following identities involving $(1,1)$-extensors: $\det[t]\tau = \underline{t}(\tau)$ and $t^{-1}(a) = \det^{-1}[t]\underline{t}^\dagger(a\tau)\tau^{-1}$, (recall that $\boldsymbol{h}^{\clubsuit} = (\boldsymbol{h}^\dagger)^{-1} = (\boldsymbol{h}^{-1})^\dagger$).

### *3.3.2 The Variational Operator* $\boldsymbol{\delta}_t^w$

Let $\mathcal{F}_{(X^1,\ldots,X^k)}[t]$ be a scalar functional of the extensor variable $t : \bigwedge^1 V \to \bigwedge^1 V$. Let also $w : \bigwedge^1 V \to \bigwedge^1 V$. Construct next the real variable function $f_w(\lambda) = \mathcal{F}_{(X^1,\ldots,X^k)}[t + \lambda w]$. Then, according to the mean value theorem we have

$$f_w(\lambda) = f_w(0) + \frac{d}{d\lambda} f_w(\lambda)|_{\lambda=0} \lambda + \frac{1}{2!}\frac{d^2}{d\lambda^2} f_w(\lambda)|_{\lambda=\lambda_1} \lambda^2, \qquad 0 < |\lambda_1| < |\lambda|. \tag{3.59}$$

The functional variation of $\mathcal{F}_{(X^1,\ldots,X^k)}[t]$ is by definition:

$$\boldsymbol{\delta}_t^w \mathcal{F}_{(X^1,\ldots,X^k)}[t] := \frac{d}{d\lambda} \mathcal{F}_w(\lambda)|_{\lambda=0}. \tag{3.60}$$

Since the Lagrangian for the field theory of gravitation (to be developed in Section 5) is a scalar functional of a $(1,1)$-extensor field $\boldsymbol{h}$ , its functional variation will play an important role in this paper. In particular, we shall need the following result.

**Example 3.11** Calculate $\boldsymbol{\delta}_t^w \det[t]$. We first construct the real variable function

$$f_w(\lambda) = \det[t + \lambda w] \tag{3.61}$$

Next, we define $g = tt^+$ and take a pair of reciprocal bases $(\{\varepsilon^i\}, \{\varepsilon_i\})$ for $V$ satisfying the condition

$$g(\varepsilon^i) \cdot \varepsilon^j = \delta^{ij}, \tag{3.62}$$

which implies that

$$t(\varepsilon^i) \cdot t(\varepsilon^j) = \delta^{ij}. \tag{3.63}$$

Recalling Eq.(2.50a), we can write

$$f_w(\lambda) = \frac{1}{n!}\{(t(\varepsilon^1) + \lambda w(\varepsilon^1)] \wedge \ldots \wedge [t(\varepsilon^n) + \lambda w(\varepsilon^n)]\}\lrcorner(\varepsilon_1 \wedge \ldots \wedge \varepsilon_n). \tag{3.64}$$

Then, we have

$$\begin{aligned} \frac{d}{d\lambda} f_w(\lambda)|_{\lambda=0} = \frac{1}{n!}\{&[w(\varepsilon^1) \wedge t(\varepsilon^2) \wedge \ldots \wedge t(\varepsilon^n)]\lrcorner(\varepsilon_1 \wedge \ldots \wedge \varepsilon_n) \\ &+ \ldots.. + [t(\varepsilon^1) \wedge t(\varepsilon^2) \wedge \ldots \wedge w(\varepsilon^n)]\lrcorner(\varepsilon_1 \wedge \ldots \wedge \varepsilon_n)\} \end{aligned} \tag{3.65}$$

and recalling the identity giving by Eq.(2.26), which says that for any $A, B, C \in \bigwedge V$, it is $(A \wedge B)\lrcorner C$ $= A\lrcorner(B\lrcorner C)$, and Eq.(2.56) $(t^{-1}(a) = \det^{-1}[t]\underline{t}^{\dagger}(a\tau)\tau^{-1})$, we can write the second member of Eq.(3.65) as

$$\begin{aligned} &\frac{1}{n!}\{w(\varepsilon^1)\lrcorner(t(\varepsilon^2) \wedge \ldots \wedge t(\varepsilon^n))\lrcorner(\varepsilon_1 \wedge \ldots \wedge \varepsilon_n) + \ldots \\ &+ (-1)^{n-1} w(\varepsilon^n)\lrcorner(t(\varepsilon^1) \wedge \ldots \wedge t(\varepsilon^{n-1}))\lrcorner(\varepsilon_1 \wedge \ldots \wedge \varepsilon_n)\} \\ &= \frac{1}{n!}\{w(\varepsilon^1)\lrcorner[t(\varepsilon^1)\lrcorner(t(\varepsilon^1)(t(\varepsilon^2) \wedge \ldots \wedge t(\varepsilon^n))\lrcorner(\varepsilon_1 \wedge \ldots \wedge \varepsilon_n)\ldots \\ &+ w(\varepsilon^n)\lrcorner[t(\varepsilon^n)\lrcorner(t(\varepsilon^1)(t(\varepsilon^2) \wedge \ldots \wedge t(\varepsilon^n))\lrcorner(\varepsilon_1 \wedge \ldots \wedge \varepsilon_n)\} \\ &= \sum_{i=1}^{n} w(\varepsilon^i)\lrcorner[t(\varepsilon^i)\lrcorner(t(\varepsilon^1)(t(\varepsilon^2) \wedge \ldots \wedge t(\varepsilon^n))\lrcorner(\varepsilon_1 \wedge \ldots \wedge \varepsilon_n) \\ &= \sum_{i=1}^{n} w(\varepsilon^i)\lrcorner t(\varepsilon^i\tau)\tau^{-1} \\ &= \sum_{i=1}^{n} w(\varepsilon^i)\lrcorner t^{\clubsuit}(\varepsilon^i) \det[t] \\ &= w(\partial_a)\lrcorner t^{\clubsuit}(a) \det[t]. \end{aligned} \tag{3.66}$$

Moreover, taking into account that, for any $a \in \bigwedge^1 V$,

$$\sum_{j=1}^{n} w(\varepsilon^i)\lrcorner t^{\clubsuit}(\varepsilon^i) \det[t] = w(\partial_a)\lrcorner t^{\clubsuit}(a) \det[t], \tag{3.67}$$

we finally get

$$\boldsymbol{\delta}_t^w \det[t] = w(\partial_a)\lrcorner t^{\clubsuit}(a) \det[t]. \tag{3.68}$$

# References

[1] Moya, . M., Fernández, V. V., and Rodrigues, W. A. Jr., Multivector Functionals, *Adv. Appl. Clifford Algebras* **11**(S3), 93-103 (2001). [http://www.clifford-algebras.org/]

# 4
# Multiform and Extensor Calculus on Manifolds

ABSTRACT In this chapter, we will introduce multiform and extensor fields on an arbitrary manifold $M$. In order to use the full algebraic machinery presented in the previous chapters, we recall from Chapter 1 that a manifold $M$ may in general support many different connections (and many different metrics). Then, given an open set $\mathcal{U} \subset M$ and chart $(\mathcal{U}, \phi)$ and a fixed point $O \in \mathcal{U}$, we define in an appropriate way a teleparallel connection[1] on $\mathcal{U} \subset M$, which permits us to construct a vector space $\mathbf{U}$ and its dual $U$, called the *canonical vector space*. This permits us to introduce on $U$, in a thoughtful way, different parallelism structures which are in a precise sense the representatives on $U$ of the restriction to $\mathcal{U}$ of parallelism structures defined by corresponding connections defined on $M$. The main object in the construction of a parallelism structure on $U$ is a connection 2-*extensor field* (and some other associated extensor fields) which permits the calculation of covariant derivative *representatives* on $U$ of multiform and extensor fields defined on $\mathcal{U} \subset M$. Moreover, given a metric structure for $M$, we introduce the concept of a metrical compatible parallelism structure ($MCPS$), present a particular $MCPS$ characterized by the Christofell operator, and introduce, moreover the 2-exform torsion field and the 4-extensor curvature field associated with a general $MCPS$ and then specialize those concepts for the case of Riemannian and Lorentzian $MCPS$. Next, we will introduce a crucial ingredient for our theory of the gravitational field, namely the concept of *elastic* and *plastic* deformations of a $MCPS$ into a new one metrical compatible parallelism structure generated by a $(1, 1)$-extensor field $\boldsymbol{h}$ that transforms the metric extensor field of the first structure into the metric extensor field of the second structure. We will study the conditions that $\boldsymbol{h}$ must satisfy in order to generate an elastic or plastic deformation. We will prove some key theorems which relate the torsion and curvature extensor fields of two structures, in which one is the deformation of the other. Particularly important for our purposes are the *gauge fields*, associated with what we call a Lorentz-Cartan metric compatible structure, which permits us to interpret a Lorentz metric compatible structure as a plastic $\boldsymbol{h}$-deformation of what we

[1]This connection on $\mathcal{U}$ cannot be extended in general to all $M$.

V.V. Fernández and W.A. Rodrigues Jr., *Gravitation as a Plastic Distortion of the Lorentz Vacuum*, Fundamental Theories of Physics 168, DOI 10.1007/978-3-642-13589-7_4,

call a Minkowski-Cartan parallelism structure. All concepts have been presented with enough details in order to help the reader become conversant with the subject.

## 4.1 Canonical Space

Let $M$ be a smooth (i.e., $C^\infty$) differential manifold, $\dim M = n$. Take an arbitrary point $O \in M$ , and a local chart $(\mathcal{U}, \phi)_o$ of an atlas of $M$, such that $O \in \mathcal{U}$.

Any point $p \in \mathcal{U}$ is then localized by an $n$-uple of real numbers $\phi(p) \in \mathbb{R}^n$, say $\phi(p) = (\xi^1(p), \ldots, \xi^n(p))$. As usual, both $\mathcal{U} \ni p \mapsto \boldsymbol{\xi}^\mu(p) \in \mathbb{R}$, the $\mu$-th coordinate function as well $\boldsymbol{\xi}^\mu(p) = \xi^\mu$, the $\mu$-th coordinate , are denoted (when no confusion arises) by the same notation $\xi^\mu$. Moreover, here $\mu = 1, \ldots, n$.

At $p \in \mathcal{U}$, the set of coordinate tangent vectors $\left\{ \left.\frac{\partial}{\partial \xi^1}\right|_{(p)}, \ldots, \left.\frac{\partial}{\partial \xi^n}\right|_{(p)} \right\}$ is a natural basis for the tangent space $T_pM$, and the set of tangent coordinate 1-forms $\left\{ \left. d\xi^1\right|_{(p)}, \ldots, \left. d\xi^n\right|_{(p)} \right\}$ is a natural basis for the cotangent (or dual) space $T_p^*M$.

We will introduce an equivalence relation on the (sub)tangent bundle $T\mathcal{U} = \bigcup_{p \in \mathcal{U}} T_pM$, as follows. Let $\mathbf{v}_p \in T_pM$ and $\mathbf{v}_q \in T_qM$. We say that $\mathbf{v}_p \sim \mathbf{v}_q$ iff $\mathbf{v}_p \xi^\mu = \mathbf{v}_q \xi^\mu$, with $\mu = 1, \ldots, n$.

This equivalence relation is well defined and, of course, it is not void, since the coordinate tangent vectors at any two points, $p, q \in \mathcal{U}$ are equivalent. Indeed, we have

$$\left.\frac{\partial}{\partial \xi^\alpha}\right|_{(p)} \xi^\mu = \left.\frac{\partial \xi^\mu}{\partial \xi^\alpha}\right|_{(p)} = \delta^\mu_\alpha = \left.\frac{\partial \xi^\mu}{\partial \xi^\alpha}\right|_{(q)} = \left.\frac{\partial}{\partial \xi^\alpha}\right|_{(q)} \xi^\mu, \quad \mu, \alpha = 1, \ldots, n, \tag{4.1}$$

and thus $\left.\frac{\partial}{\partial \xi^\alpha}\right|_{(p)} \sim \left.\frac{\partial}{\partial \xi^\alpha}\right|_{(q)}$.

For the point $O \in \mathcal{U}$, the equivalence class of a tangent vector $\mathbf{v}_o \in T_oM$ is given by $[\mathbf{v}_o] = \{\mathbf{v}_p \sim \mathbf{v}_o \mid p \in \mathcal{U}\}$.

The set of the equivalence classes of each one of the tangent vectors $\mathbf{v}_o \in T_oM$, i.e., $\mathbf{U} = \{[\mathbf{v}_o] \mathbin{/} \mathbf{v}_o \in T_oM\}$, equipped with the *sum* of equivalence classes and the *multiplication* of equivalence classes by scalars (real numbers), defined by

$$[\mathbf{v}_o] + [\mathbf{w}_o] = [\mathbf{v}_o + \mathbf{w}_o]; \qquad \alpha[\mathbf{v}_o] = [\alpha \mathbf{v}_o], \tag{4.2}$$

has a natural structure of vector space over $\mathbb{R}$.

The equivalence class of the null vector $\mathbf{0}_o \in T_oM$, i.e., $\mathbf{0} \equiv [\mathbf{0}_o] \in \mathbf{U}$, i.e., the set of all null vectors in $T\mathcal{U}$ is the null vector of $\mathbf{U}$.

The set of the equivalence classes of the $n$-tangent coordinate vectors $\left\{ \left.\frac{\partial}{\partial \xi^i}\right|_{(o)} \right\}$, with $\left.\frac{\partial}{\partial \xi^i}\right|_{(o)} \in T_oM$, say $\{\mathbf{b}_1, \ldots, \mathbf{b}_n\}$ with

$$\mathbf{b}_1 \equiv [\left.\frac{\partial}{\partial \xi^1}\right|_{(o)}], \ldots, \mathbf{b}_n \equiv [\left.\frac{\partial}{\partial \xi^n}\right|_{(o)}] \in \mathbf{U}, \tag{4.3}$$

is a basis for $\mathbf{U}$, called the *fiducial basis* and, of course, $\dim \mathbf{U} = n$.

$\mathbf{U}$ is a vector space, obviously associated with the chart $(\mathcal{U}, \phi)_o$, but nevertheless it will be said to be the *canonical space*, since it will play a fundamental role in our theory of multiform and extensor fields on manifolds. To continue, we recall that from the above identifications, we have immediately the identification of $\bigwedge T_p^* M$ with $\bigwedge T_q^* M$ for all $p, q \in \mathcal{U}$.

Taking into account the notations introduced in Chapter 2 the dual space of $\mathbf{U}$ is denoted by $U$, the space of $k$-forms is denoted by $\bigwedge^k U$ and the space of multiforms is denoted by $\bigwedge U$.

The dual basis of the fiducial basis $\{\mathbf{b}_\mu\}$ is said to be a fiducial basis for $U$ and is denoted by $\{\boldsymbol{\beta}^\mu\}$, i.e., $\boldsymbol{\beta}^\mu(\mathbf{b}_\nu) = \delta^\mu_\nu$. To be able to use Einstein's sum convention we will introduce also the notation,

$$b^\mu = \mathbf{b}_\mu \; e \; \beta_\mu = \boldsymbol{\beta}^\mu. \tag{4.4}$$

The canonical scalar product is constructed as in Chapter 2, using $\{\mathbf{b}_\mu\}$, and in what follows, we use for our calculations the canonical Clifford algebra $\mathcal{C}\ell(U, \cdot)$.

The Position 1-Form

To any $p \in \mathcal{U}$ there correspond exactly $n$ real numbers $\xi^1, \ldots, \xi^n$, its position coordinates in the local chart $(\mathcal{U}, \phi)_o$. It is thus possible to define a 1-form on $U$,

$$x = \xi^\mu \beta_\mu, \tag{4.5}$$

associated with $p \in \mathcal{U}$. The 1-form $x \in U$ *localizes* $p$ in the vector space $U$ and we call $x$ the *position 1-form (or position vector) of* $p$.

Observe that given an arbitrary 1-form on $U$, it will not necessarily be the position vector of some point of the open set $\mathcal{U}$, since the components of an arbitrary element of $U$ do not need to be the coordinates of some point of $\mathcal{U}$ necessarily.

The set of all position vectors of the points of $\mathcal{U}$ is denoted $U_o \subset U$. This subset of the vector space $U$ is not necessarily a vector subspace of $U$.

Observe also that we have immediately, for any $x, y \in U_o$, the identification of the tangent spaces $\bigwedge T_x^* U$ and $\bigwedge T_y^* U$ with $\bigwedge U$.

### 4.1.1 Multiform Fields

From what has been said, it is obvious that any $\mathcal{X} \in \sec \bigwedge T^*M$, when restricted to $\mathcal{U}$, will be represented by a multiform function $X$ of the position vector, which we will write as

$$X \in \sec \bigwedge T^* U, \tag{4.6}$$

meaning that for each $x \in U_o$, $X(x) = X_{(x)} \in \bigwedge T_x^* U \simeq \bigwedge U$. Eventually, we also use the notation $X : U_0 \to \bigwedge U$.

If $X \in \sec \bigwedge T^* U$ is $C^\infty$-differentiable on the set $U_0$, then the multiform $\mathcal{X} \in \sec \bigwedge T^*M$ is differentiable on $\mathcal{U}$.

Extensor Fields

Any $(p, q)$-extensor field $\mathfrak{t}$ on $M$, when restricted to $\mathcal{U}$, is represented by an extensor function $t$ of the position vector. We write

$$t \in \sec(p, q)\text{-}ext U, \tag{4.7}$$

and eventually also write $t : U_0 \to (p,q)\text{-}extU$.

If $t$ is $C^\infty$-differentiable[2] on $U_o$, then the $(p,q)$-extensor field $\mathfrak{t}$ is smooth on $\mathcal{U}$. For a general extensor field, say $\Delta$, we use the notation $\Delta \in \sec extU$.

## 4.2 Parallelism Structure $(U_0, \gamma)$ and Covariant Derivatives

Given a smooth manifold $M$ ($\dim M = n$) and an *arbitrary* linear connection $\nabla$ on $M$, the pair $(M, \nabla)$ is said to be a *parallelism structure*. We will now analyze some important parallelism structures that will appear in the next sections.

### 4.2.1 The Connection 2-Extensor Field $\gamma$ on $U_o$ and Associated Extensor Fields

The parallelism structure $(M, \nabla)$ is represented on $U_o$ by a pair $(U_0, \gamma)$ (also called a parallelism structure) where $\gamma \in \sec extU$ is called a smooth *connection 2-extensor field* on $U_o$, i.e., for each $x \in U_0$,

$$\gamma_{(x)} : \bigwedge^1 T_xU \times \bigwedge^1 T_xU \to \bigwedge^1 T_xU. \tag{4.8}$$

Also, we will define the following fields associated with $\gamma$. First, given $a \in \sec \bigwedge^1 T^*U$, define the smooth $(1,1)$-extensor field on $U_o$, $\gamma_a \in \sec(1,1)\text{-}extU$ by

$$\gamma_a(b) = \gamma(a, b) \tag{4.9}$$

for all $b \in \sec \bigwedge^1 T^*U$. The field $\gamma_a$ is called a *connection* $(1,1)$-*extensor field on* $U_o$.

Next, we introduce a smooth $(1,2)$-extensor field on $U_0$, say $\omega \in \sec(1,2)\text{-}extU$ such that for $a \in \sec \bigwedge^1 T^*U$, $\omega(a) \in \sec \bigwedge^2 T^*U$ is given by

$$\omega(a) := \frac{1}{2}\mathrm{bif}[\gamma_a]. \tag{4.10}$$

The field $\omega$ will be called (for reasons that will become clear in a while) a *rotation gauge field.*

Finally we define the smooth extensor field on $U_o$, $\Gamma_a \in \sec extU$ such that for each $x \in U_o$, $\Gamma_{(x)a}$ is the *generalized* of $\gamma_{(x)a}$ (recall Eq.(2.61)).

### 4.2.2 Covariant Derivative of Multiform Fields Associated with $(U_0, \gamma)$

The connection 2-extensor field $\gamma$ is used in the following way. For any smooth $a \in \sec \bigwedge^1 T^*U$ which is a representative on $U$ of a vector field $\mathfrak{a} \in \sec T\mathcal{U}$ we will introduce two covariant operators $\nabla_a$ and $\nabla_a^-$, acting on the module of smooth multiform fields on $U_o$.

First, if $b \in \sec \bigwedge^1 T^*U$

$$\nabla_a b := a \cdot \partial b + \gamma_a(b), \tag{4.11a}$$

$$\nabla_a^- b := a \cdot \partial b - \gamma_a(b) \tag{4.11b}$$

[2] A $(p,q)$-extensorial function $t$ is said to be $C^\infty$-differentiable on $U_0$ iff for an any $C^\infty$-differentiable $p$-form function on $U_0$, say $x \mapsto X(x)$, the $q$-form function $x \mapsto t_{(x)}(X(x))$ is $C^\infty$-differentiable on $U_0$.

For any smooth $X \in \sec \bigwedge T^*U$ we define

$$\nabla_a X := a \cdot \partial X + \Gamma_a(X), \tag{4.12a}$$

$$\nabla_a^- X := a \cdot \partial X - \Gamma_a^\dagger(X), \tag{4.12b}$$

where $\Gamma_a$ is the generalized of $\gamma_a$ and $\Gamma_a^\dagger$ is the adjoint of $\Gamma_a$.

The operators $\nabla_a$ and $\nabla_a^-$ are well defined covariant derivatives on $U_o$ and thus satisfy all properties of a covariant derivative operator, i.e., they are linear with respect to the direction 1-form field, i.e., we have

$$\nabla^{\pm}_{\alpha a+\beta b} X = \alpha \nabla^{\pm}_a X + \beta \nabla^{\pm}_b X, \tag{4.13}$$

for $\alpha, \beta \in \mathbb{R}$ and $a, b \in \sec \bigwedge^1 T^*U$ and

$$\nabla_a f = a \cdot \partial f, \;\; \nabla_a(X+Y) = \nabla_a X + \nabla_a Y, \;\; \nabla_a(fX) = a \cdot \partial f + f\nabla_a X, \tag{4.14}$$

with analog equations for $\nabla_a^-$.

To prove, e.g., that $\nabla_a f = a \cdot \partial f$, it is enough to recall a property of the generalized $T$ of a given $(1,1)$-extensor $t$, explicitly $T(\alpha) = 0$, with $\alpha \in \mathbb{R}$. Then, we have, using Eq.(4.12a) that for any smooth scalar field $f$ indeed $\nabla_a f = a \cdot \partial f + \Gamma_a(f) = a \cdot \partial f$.

The proofs of the remaining formulas in Eq.(4.14) are equally elementary.

We can prove also that $\nabla_a$ and $\nabla_a^-$ satisfy *Leibniz rule* for the exterior product of smooth multiform fields, i.e.,

$$\nabla_a(X \wedge Y) = (\nabla_a X) \wedge Y + X \wedge (\nabla_a Y), \tag{4.15}$$

with an analogous formula for $\nabla_a^-$. Indeed, take two smooth multiform fields $X$ and $Y$. Using the Leibniz rule, satisfied by the directional derivative operator $a \cdot \partial$ when applied to the exterior product of smooth multiform fields and a property of the generalized of a $(1,1)$-extensor $t$ (namely $T(A \wedge B) = T(A) \wedge B + A \wedge T(B)$, for any $A, B \in \sec \bigwedge T^*U$), we get

$$\begin{aligned} \nabla_a(X \wedge Y) &= a \cdot \partial(X \wedge Y) + \Gamma_a(X \wedge Y) \\ &= (a \cdot \partial X) \wedge Y + X \wedge (a \cdot \partial Y) + \Gamma_a(X) \wedge Y + X \wedge \Gamma_a(Y) \\ \nabla_a(X \wedge Y) &= (\nabla_a X) \wedge Y + X \wedge (\nabla_a Y), \end{aligned} \tag{4.16}$$

which proves Eq.(4.15).

On the other hand, the operator of ordinary directional derivative $a \cdot \partial$ and the operators $\nabla_a$ and $\nabla_a^-$ are related by a notable identity:

$$a \cdot \partial(X \cdot Y) = (\nabla_a X) \cdot Y + X \cdot (\nabla_a^- Y) = (\nabla_a^- X) \cdot Y + X \cdot (\nabla_a Y). \tag{4.17}$$

Indeed, take two smooth multiform fields $X$ and $Y$. Using the Leibniz rule for $a \cdot \partial$, when applied to the exterior product of smooth multiform fields and that $T(A) \cdot B = A \cdot T^\dagger(B)$, for all $A, B \in \sec \bigwedge T^*U$, we see immediately the validity of the formulas

$$\begin{aligned} (\nabla_a X) \cdot Y + X \cdot (\nabla_a^- Y) &= (a \cdot \partial X) \cdot Y + \Gamma_a(X) \cdot Y \\ &\quad + X \cdot (a \cdot \partial Y) - X \cdot \Gamma_a^\dagger(Y) \\ &= (a \cdot \partial X) \cdot Y + X \cdot (a \cdot \partial Y) \\ (\nabla_a X) \cdot Y + X \cdot (\nabla_a^- Y) &= a \cdot \partial(X \cdot Y). \end{aligned} \tag{4.18}$$

### *4.2.3 Covariant Derivative of Extensor Fields Associated with $(U_0, \gamma)$*

The covariant derivative operators $\nabla_a$ and $\underset{a}{\nabla^-}$ may be extended to act on the *module* of smooth $(p, q)$-extensor fields on $U_o$ (which represent smooth $(p, q)$-extensor field on $\mathcal{U} \subset M$). We define for $t \in \sec(p, q)$-*extU*, $\nabla_a t$ and $\underset{a}{\nabla^-} t$ as the smooth $(p, q)$-extensor fields on $U_o$, such that for any smooth $X \in \sec \bigwedge^p T^*U$,

$$(\nabla_a t)(X) := \nabla_a t(X) - t(\underset{a}{\nabla^-} X), \tag{4.19a}$$
$$(\underset{a}{\nabla^-} t)(X) := \underset{a}{\nabla^-} t(X) - t(\nabla_a X). \tag{4.19b}$$

Observe that, in the above formulas $\nabla_a t(X)$ denotes the covariant derivative $\nabla_a$ of the smooth multiform field $t(X)$ and $\underset{a}{\nabla^-} X$ is the covariant derivative $\underset{a}{\nabla^-}$ of the smooth multiform field $X$.

Certainly, those properties are consistent with the linearity property of the smooth $(p, q)$-extensor fields. Indeed, take two smooth $p$-form fields $X$ and $Y$, and a smooth scalar field $f$. We have immediately

$$\begin{aligned}(\nabla_a t)(X + Y) &= \nabla_a t(X + Y) - t(\underset{a}{\nabla^-}(X + Y)) \\ &= \nabla_a(t(X) + t(Y)) - t(\underset{a}{\nabla^-} X + \underset{a}{\nabla^-} Y) \\ &= \nabla_a t(X) + \nabla_a t(Y) - t(\underset{a}{\nabla^-} X) - t(\underset{a}{\nabla^-} Y) \\ &= (\nabla_a t)(X) + (\nabla_a t)(Y),\end{aligned} \tag{4.20}$$

and

$$\begin{aligned}(\nabla_a t)(fX) &= \nabla_a t(fX) - t(\underset{a}{\nabla^-}(fX)) \\ &= \nabla_a(ft(X)) - t((a \cdot \partial f)X + f\underset{a}{\nabla^-} X) \\ &= (a \cdot \partial f)t(X) + f\nabla_a t(X) - (a \cdot \partial f)t(X) - ft(\underset{a}{\nabla^-} X) \\ &= f(\nabla_a t)(X).\end{aligned} \tag{4.21}$$

Each one of the covariant derivatives $t \mapsto \nabla_a t$ and $t \mapsto \underset{a}{\nabla^-} t$ possesses the property of linearity with relation to the direction 1-form, i.e., $\underset{\alpha a+\beta b}{\nabla^{\pm}} t = \alpha \underset{a}{\nabla^{\pm}} t + \beta \underset{b}{\nabla^{\pm}} t$, where $\alpha, \beta \in \mathbb{R}$ and $a, b \in \sec \bigwedge^1 T^*U$

Moreover, $t \mapsto \nabla_a t$,and $t \mapsto \underset{a}{\nabla^-} t$ possess also the following properties:

$$\underset{a}{\nabla^{\pm}}(t + u) = \underset{a}{\nabla^{\pm}} t + \underset{a}{\nabla^{\pm}} u, \tag{4.22}$$
$$\underset{a}{\nabla^{\pm}}(ft) = (a \cdot \partial f)t + f\underset{a}{\nabla^{\pm}} t, \tag{4.23}$$

where $t$ and $u$ are smooth $(p, q)$-extensor fields and $f$ is a scalar field.

Let us prove Eq.(4.22).and Eq.(4.23). Let $X$ be an arbitrary smooth $p$-form field,then

$$\begin{aligned}(\nabla_a(t + u))(X) &= \nabla_a(t + u)(X) - (t + u)(\underset{a}{\nabla^-} X) \\ &= \nabla_a(t(X) + u(X)) - t(\underset{a}{\nabla^-} X) - u(\underset{a}{\nabla^-} X) \\ &= \nabla_a t(X) + \nabla_a u(X) - t(\underset{a}{\nabla^-} X) - u(\underset{a}{\nabla^-} X) \\ &= (\nabla_a t)(X) + (\nabla_a u)(X) = (\nabla_a t + \nabla_a u)(X),\end{aligned} \tag{4.24}$$

i.e., $\nabla_a(t+u) = \nabla_a t + \nabla_a u$. Also,

$$\begin{aligned}(\nabla_a(ft))(X) &= \nabla_a(ft)(X) - ft(\nabla_a X) = \nabla_a ft(X) - ft(\nabla_a X) \\ &= (a \cdot \partial f)t(X) + f\nabla_a t(X) - ft(\nabla_a X) \\ &= (a \cdot \partial f)t(X) + f(\nabla_a t)(X), \end{aligned} \tag{4.25}$$

i.e., $\nabla_a(ft) = (a \cdot \partial f)t + f(\nabla_a t)$.

### *4.2.4 Notable Identities*

We will end this section by presenting two notable identities, which are:

**(i)** Let $X$ be a smooth $p$-form field and $Y$ a smooth $q$-form field, then

$$(\nabla_a t)(X) \cdot Y = a \cdot \partial(t(X) \cdot Y) - t(\nabla_a^- X) \cdot Y - t(X) \cdot \nabla_a^- Y. \tag{4.26}$$

Indeed, we have

$$\begin{aligned}(\nabla_a t)(X) \cdot Y &= \nabla_a t(X) \cdot Y - t(\nabla_a^- X) \cdot Y \\ &= a \cdot \partial t(X) \cdot Y + \Gamma_a(t(X)) \cdot Y - t(\nabla_a^- X) \cdot Y \\ &= a \cdot \partial t(X) \cdot Y + t(X) \cdot \Gamma_a^\dagger(Y) - t(\nabla_a^- X) \cdot Y \\ &= a \cdot \partial t(X) \cdot Y + t(X) \cdot a \cdot \partial Y - t(X) \cdot a \cdot \partial Y \\ &+ t(X) \cdot \Gamma_a^\dagger(Y) - t(\nabla_a^- X) \cdot Y \\ &= a \cdot \partial t(X) \cdot Y + t(X) \cdot a \cdot \partial Y - t(\nabla_a^- X) \cdot Y \\ &- t(X) \cdot (a \cdot \partial Y - \Gamma_a^\dagger(Y)) \\ &= a \cdot \partial(t(X) \cdot Y) - t(\nabla_a^- X) \cdot Y - t(X) \cdot \nabla_a^- Y.\end{aligned}$$

**(ii)** For any smooth $(p, q)$-extensor field $t$, it is

$$\nabla_a t^\dagger = (\nabla_a t)^\dagger. \tag{4.27}$$

Take the smooth fields $X \in \sec \bigwedge^p T^*U$, and $Y \in \sec \bigwedge^q T^*U$. Utilizing Eq.(4.26) twice and the algebraic property $t(X) \cdot Y = X \cdot t^\dagger(Y)$, we have

$$\begin{aligned}(\nabla_a t^\dagger)(Y) \cdot X &= a \cdot \partial(t^\dagger(Y) \cdot X) - t^\dagger(\nabla_a^- Y) \cdot X - t^\dagger(Y) \cdot \nabla_a^- X \\ &= a \cdot \partial(t(X) \cdot Y) - t(X) \cdot \nabla_a^- Y - t(\nabla_a^- X) \cdot Y \\ &= (\nabla_a t)(X) \cdot Y = X \cdot (\nabla_a t)^\dagger(Y), \end{aligned} \tag{4.28}$$

which implies $(\nabla_a t^\dagger)(Y) = (\nabla_a t)^\dagger(Y)$, i.e., $\nabla_a t^\dagger = (\nabla_a t)^\dagger$.

For $\nabla_a^-$ completely analogous properties hold, i.e., for any smooth $p$-form field $X$ and all smooth $q$-form field $Y$, it is

$$(\nabla_a^- t)(X) \cdot Y = a \cdot \partial(t(X) \cdot Y) - t(\nabla_a X) \cdot Y - t(X) \cdot \nabla_a Y. \tag{4.29}$$

Also, for any smooth $(p, q)$-extensor field $t$,

$$\nabla_a^- t^\dagger = (\nabla_a^- t)^\dagger. \tag{4.30}$$

### *4.2.5 The 2-Exform Torsion Field of the Structure* $(U_o, \gamma)$

Given a parallelism structure $(M, \nabla)$, we know that the torsion and curvature operators (used for the definition of the torsion and Riemann curvature tensors) characterize $\nabla$ completely. Let $(U_o, \gamma)$ be the representative of the structure $(M, \nabla)$ restricted to $\mathcal{U}$. We will now introduce the *representatives* of the torsion and curvature operators of $(M, \nabla)$ on $U_o$.

First we will introduce the smooth torsion 2-*exform*[3] field $\overset{\gamma}{\tau} \in \sec extU$, such for smooth $a, b \in \sec \bigwedge^1 T^*U$ we have $(a, b) \mapsto \overset{\gamma}{\tau}(a, b) \in \sec \bigwedge^1 T^*U$, given by

$$\overset{\gamma}{\tau}(a, b) := \nabla_a b - \nabla_b a - [a, b] = \gamma_a(b) - \gamma_b(a). \tag{4.31}$$

Next we will introduce the $(2,1)$-extensor field $\overset{\gamma}{T} \in \sec(2,1)\text{-}extU$,such that, for any smooth $B \in \sec \bigwedge^1 T^*U$, $B \mapsto \overset{\gamma}{T}(B) \in \sec \bigwedge^1 T^*U$, we have:

$$\overset{\gamma}{T}(B) := \frac{1}{2} B \cdot (\partial_a \wedge \partial_b) \overset{\gamma}{\tau}(a, b). \tag{4.32}$$

$\overset{\gamma}{T}$ is called the *torsion* $(2,1)$-*extensor field.*

Observe that, if the covariant derivative $\nabla_a$ is *symmetric* (i.e., $\nabla_a b - \nabla_b a = [a, b]$, for all smooth 1-form fields $a$ and $b$, then $\overset{\gamma}{\tau}(a, b) = 0$ and $\overset{\gamma}{T}(B) = 0$.

## 4.3 Curvature Operator and Curvature Extensor Fields of the Structure $(U_o, \gamma)$

Now we will introduce a linear operator that acts on the Lie algebra of smooth 1-form fields on $U_o$. Given $a, b, c \in \sec \bigwedge^1 T^*U$ the curvature operator is the mapping

$$\begin{array}{rccc} \overset{\gamma}{\rho} : & \sec(\bigwedge^1 T^*U \times \bigwedge^1 T^*U \times \bigwedge^1 T^*U) & \to & \sec \bigwedge^1 T^*U, \\ & (a, b, c) & \mapsto & \overset{\gamma}{\rho}(a, b, c), \end{array}$$

such that

$$\overset{\gamma}{\rho}(a, b, c) := [\nabla_a, \nabla_b]c - \nabla_{[a,b]}c. \tag{4.33}$$

The operator $\overset{\gamma}{\rho}$ characterizes the curvature of the parallelism structure $(M, \nabla)$ on $U_0$. Its main properties are:

**(i)** $\overset{\gamma}{\rho}$ is antisymmetric with respect to the first and second variables, i.e.,

$$\overset{\gamma}{\rho}(a, b, c) = -\overset{\gamma}{\rho}(b, a, c). \tag{4.34}$$

**(ii)** If the covariant derivative $\nabla_a$ is symmetric, i.e., $\nabla_a b - \nabla_b a = [a, b]$ (or equivalently $\gamma_a(b) = \gamma_b(a)$, for all smooth 1-form fields $a$ and $b$) then $\overset{\gamma}{\rho}$ has a cyclic property, i.e.,

$$\overset{\gamma}{\rho}(a, b, c) + \overset{\gamma}{\rho}(b, c, a) + \overset{\gamma}{\rho}(c, a, b) = 0. \tag{4.35}$$

---

[3]A 2-exform on $U$ is an antisymmetric 2-extensor on $U$ , i.e., a linear mapping $\theta : \bigwedge^1 U \times \bigwedge^1 U \to \bigwedge U$ such that for all $a, b \in \bigwedge^1 U$ it is $\theta(a, b) = -\theta(b, a)$.

The smooth scalar 4-extensor field[4], $(w, a, b, c) \mapsto \overset{\gamma}{\mathbf{R}}_4(w, a, b, c)$, such that

$$\overset{\gamma}{\mathbf{R}}_1(w, a, b, c) = w \cdot \overset{\gamma}{\rho}(b, c, a), \tag{4.36}$$

for all smooth 1-form fields $w, a, b$ and $c$, is called the curvature 4-*extensor field*

The main properties of $\overset{\gamma}{\mathbf{R}}_1$ are:

**(i)** $\overset{\gamma}{\mathbf{R}}_1$ is antisymmetric with respect to the third and fourth variables, i.e.,

$$\overset{\gamma}{\mathbf{R}}_1(w, a, b, c) = -\overset{\gamma}{\mathbf{R}}_1(w, a, c, b). \tag{4.37}$$

**(ii)** If the covariant derivative $\nabla_a$ is symmetric then $\overset{\gamma}{\mathbf{R}}_4$ possess a cyclic property, i.e.,

$$\overset{\gamma}{\mathbf{R}}_1(w, a, b, c) + \overset{\gamma}{\mathbf{R}}_1(w, b, c, a) + \overset{\gamma}{\mathbf{R}}_1(w, c, a, b) = 0. \tag{4.38}$$

The smooth scalar 2-extensor field $(a, b) \mapsto \overset{\gamma}{\mathbf{R}}_2(a, b)$, such that

$$\overset{\gamma}{\mathbf{R}}_2(a, b) = \overset{\gamma}{\mathbf{R}}_1(\partial_w, a, w, b), \tag{4.39}$$

for all smooth 1-form fields $a$ and $b$, is called the *Ricci* 2-*extensor field.* Note that $\overset{\gamma}{\mathbf{R}}_2$ is an *internal contraction* of $\overset{\gamma}{\mathbf{R}}_1$ between the first and third variables calculated as follows. Let $(\{\varepsilon^\mu\}, \{\varepsilon_\mu\})$ be a pair of reciprocal bases on $U_0$, then an internal contraction between the first and third variables is

$$\overset{\gamma}{\mathbf{R}}_1(\partial_w, a, w, b) := \frac{\partial}{\partial w_\mu}\overset{\gamma}{\mathbf{R}}_1(\varepsilon^\mu, a, w, b) = \overset{\gamma}{\mathbf{R}}_1(\varepsilon^\mu, a, \varepsilon_\mu, b) = \overset{\gamma}{\mathbf{R}}_1(\varepsilon_\mu, a, \varepsilon^\mu, b),$$

with $\mu$ summed from 0 to 3.

The smooth $(1,1)$-extensor field $b \mapsto \overset{\gamma}{\boldsymbol{R}}_1(b)$ such that

$$a \cdot \overset{\gamma}{\boldsymbol{R}}_1(b) = \overset{\gamma}{\mathbf{R}}_2(a, b), \tag{4.40}$$

for all smooth 1-form fields $a$ and $b$, is called the *Ricci* $(1,1)$-*extensor field.*

Note that we can write $\overset{\gamma}{\boldsymbol{R}}_1(b)$ as

$$\overset{\gamma}{\boldsymbol{R}}_1(b) = \partial_a \overset{\gamma}{\mathbf{R}}_2(a, b), \tag{4.41}$$

once we recall that $\partial_a \overset{\gamma}{\mathbf{R}}_2(a, b) \equiv \varepsilon^\mu(\varepsilon_\mu \cdot \partial_a \overset{\gamma}{\mathbf{R}}_2(a, b))$, i.e., $\partial_a \overset{\gamma}{\mathbf{R}}_2(a, b) = \varepsilon^\mu \overset{\gamma}{\mathbf{R}}_2(\varepsilon_\mu, b)$ due to the linearity of $\overset{\gamma}{\mathbf{R}}_2$ with respect to the first variable.

---

[4]A linear mapping $t : \sec(\bigwedge^1 U \times \bigwedge^1 U \times \bigwedge^1 U \times \bigwedge^1 U) \to \mathbb{R}$ is called a scalar 4-extensor.

## 4.4 Covariant Derivatives Associated with Metric Structures $(U_o, \boldsymbol{g})$

### *4.4.1 Metric Structures*

Let $\mathbf{g}$ be a smooth $(1,1)$-extensor field (associated with a metric tensor $\mathbf{g} \in \sec T_2^0 M$ ). Its representative $\boldsymbol{g}$ on a given $U_o$ is symmetric (i.e., $\boldsymbol{g}_{(x)} = \boldsymbol{g}_{(x)}^{\dagger}$ for all $x \in U_0$), is non degenerate (i.e., $\det[\boldsymbol{g}_{(x)}] \neq 0$ for $x \in U_0$). In what follows, we say that $\boldsymbol{g}$ is a metric extensor field on $U_0$.

A pair $(M, \mathbf{g})$ is said to be a *metric structure* and its restriction to $\mathcal{U} \subset M$ is represented by $(U_o, \boldsymbol{g})$, and is also called a metric structure.

### *4.4.2 Christoffel Operators for the Metric Structure $(U_o, \mathbf{g})$*

Let $\boldsymbol{g}$ be the representative of the extensor field $\mathbf{g}$ (associated to $\mathbf{g}$) on $U_o$.

The two classical Christoffel operators on $\mathcal{U} \subset M$ are represented on the structure $(U_o, \boldsymbol{g})$ by two operators that have the same name and that act on the Lie algebra of smooth 1-form fields on $U_0$. If $a, b, c \in \sec \bigwedge^1 T^*U$ are such fields, we have

**(i)** *The first Christoffel operator* is the smooth mapping $(a, b, c) \mapsto [a, b, c]$ such that

$$[a,b,c] := \frac{1}{2}(a \cdot \partial(\boldsymbol{g}(b) \cdot c) + b \cdot \partial(\boldsymbol{g}(c) \cdot a) - c \cdot \partial(\boldsymbol{g}(a) \cdot b) + \boldsymbol{g}(c) \cdot [a,b] + \boldsymbol{g}(b) \cdot [c,a] - \boldsymbol{g}(a) \cdot [b,c]), \tag{4.42}$$

where $[a, b]$ is the Lie bracket of $a$ and $b$ defined by:

$$[a,b] := a \cdot \partial b - b \cdot \partial a. \tag{4.43}$$

**(ii)** *The second Christoffel operator,* $(a, b, c) \mapsto \left\{ \begin{matrix} c \\ a, b \end{matrix} \right\}$ is defined by

$$\left\{ \begin{matrix} c \\ a, b \end{matrix} \right\} = [a, b, \boldsymbol{g}^{-1}(c)]. \tag{4.44}$$

Note that $\left\{ \begin{matrix} c \\ a, b \end{matrix} \right\}$ is also a smooth 1-form field, since has been defined algebraically from $[a, b, c]$.

For all smooth form fields $a, a', b, b', c, c'$ and smooth scalar field $f$, the first Christoffel operator satisfies the elementary properties:

$$\begin{aligned}
[a + a', b, c] &= [a, b, c] + [a', b, c]. \\
[fa, b, c] &= f[a, b, c]. \\
[a, b + b', c] &= [a, b, c] + [a, b', c]. \\
[a, fb, c] &= f[a, b, c] + (a \cdot \partial f)\boldsymbol{g}(b) \cdot c. \\
[a, b, c + c'] &= [a, b, c] + [a, b, c']. \\
[a, b, fc] &= f[a, b, c].
\end{aligned} \tag{4.45}$$

Eq.(4.45) says that $[a, b, c]$ is linear with respect to the first and third arguments, but it is *not* linear with respect to the second argument.

Given the smooth 1-form fields $a, b$ and $c$, other relevant properties of $[a, b, c]$ are:

$$\begin{aligned}
[a,b,c]+[b,a,c] &= a\cdot\partial(\boldsymbol{g}(b)\cdot c)+b\cdot\partial(\boldsymbol{g}(c)\cdot a)-c\cdot\partial(\boldsymbol{g}(a)\cdot b)\\
&\quad+\boldsymbol{g}(b)\cdot[c,a]-\boldsymbol{g}(a)\cdot[b,c] &(4.46a)\\
[a,b,c]-[b,a,c] &= \boldsymbol{g}(c)\cdot[a,b]. &(4.46b)\\
[a,b,c]+[a,c,b] &= a\cdot\partial(\boldsymbol{g}(b)\cdot c) &(4.46c)\\
[a,b,c]-[a,c,b] &= b\cdot\partial(\boldsymbol{g}(c)\cdot a)-c\cdot\partial(\boldsymbol{g}(a)\cdot b)+\boldsymbol{g}(c)\cdot[a,b]\\
&\quad+\boldsymbol{g}(b)\cdot[c,a]-\boldsymbol{g}(a)\cdot[b,c]. &(4.46d)\\
[a,b,c]+[c,b,a] &= b\cdot\partial(\boldsymbol{g}(c)\cdot a)+\boldsymbol{g}(c)\cdot[a,b]-\boldsymbol{g}(a)\cdot[b,c] &(4.46e)\\
[a,b,c]-[c,b,a] &= a\cdot\partial(\boldsymbol{g}(b)\cdot c)-c\cdot\partial(\boldsymbol{g}(a)\cdot b)+\boldsymbol{g}(b)\cdot[c,a]. &(4.46f)
\end{aligned}$$

### *4.4.3 The 2-Extensor field* $\lambda$

Given the metric structure $(M, \boldsymbol{g})$ and $(U_o, \boldsymbol{g})$, the representative on $U_o$ of its restriction to $\mathcal{U}$, we can construct a connection 2-extensor field $\lambda$ on $U_0$, $\lambda \in \sec ext U$, such that $a, b \in \sec \bigwedge^1 T^*U$,

$$\lambda : \sec(\bigwedge^1 T^*U \times \bigwedge^1 T^*U) \to \sec \bigwedge^1 T^*U,$$

$$\lambda(a,b) = \partial_c \left\{ \begin{matrix} c \\ a,b \end{matrix} \right\} - a \cdot \partial b. \tag{4.47}$$

**Remark 4.1** In complete analogy to the case of connection 2-extensor field $\gamma$, we will also introduce here the extensor fields $\lambda_a$ and $\Lambda_a$. Those objects and some others constructed from them (see next subsection), enter in the calculations of covariant derivatives of multiform fields and extensor fields defined on $U_o$. The field $\lambda$ will be called a *Levi-Civita connection* on $U_o$.

### *4.4.4 (Riemann and Lorentz)-Cartan MGSS's* $(U_o, \boldsymbol{g}, \gamma)$

Consider a triple $(M, \mathbf{g}, \nabla)$ where $M$ is a smooth manifold ($\dim M = n$), $\mathbf{g} \in \sec T_2^0 M$ is a Riemannian or a Lorentzian metric tensor, and $\nabla$ is an arbitrary *metric compatible connection.* Let $\mathbf{g}$ be the metric extensor corresponding to $\mathbf{g}$ and let $\boldsymbol{g}$ be the representative of $\mathbf{g}$ on $U_o$. As we already know, the connection $\nabla$ is characterized on $U_o$ by a smooth 2-extensor field $\gamma$ (and other two extensor fields defined from $\gamma$). The metric compatibility means

$$\nabla \mathbf{g} = 0, \tag{4.48}$$

and such condition is obviously expressed on $U_0$ by

$$\nabla_a^- \boldsymbol{g} = 0. \tag{4.49}$$

In what follows, when $\mathbf{g}$ has Euclidean signature ($(n, 0)$ or $(0, n)$), we call $(M, \mathbf{g}, \nabla)$ a *Riemann-Cartan MCGSS. When* $\mathbf{g}$ has pseudo-euclidean signature $(1, n-1)$, we call $(M, \mathbf{g}, \nabla)$ a *Lorentz-Cartan MCGSS*[5]. We use also such denominations for the triple $(U_o, \boldsymbol{g}, \gamma)$.

[5]The *GSS* where $\boldsymbol{g}$ has other possible signatures may be called semi-riemannian.

### 4.4.5 *Existence Theorem of the $\gamma\boldsymbol{g}$-gauge Rotation Extensor of the MCGSS* $(U_o, \boldsymbol{g}, \gamma)$

Now we will present a theorem whose proof may be found in [1], and which plays a crucial role in our theory of the gravitational field in Chapter 6.

**Theorem 4.1** [2] On the *MCGSS* $(U_o, \boldsymbol{g}, \gamma)$, there exists a $(1,2)$-extensor field $\overset{\gamma g}{\omega}(a) = \frac{1}{2}\overset{g}{\text{bif}}[\gamma_a]$, such that for all smooth 1-form fields $a$ and $b$

$$\gamma_a(b) = \frac{1}{2}\boldsymbol{g}^{-1}(a \cdot \partial\boldsymbol{g})(b) + \overset{\gamma g}{\omega}(a) \underset{g}{\times} b, \tag{4.50}$$

where

$$\overset{\gamma g}{\omega}(a) \underset{g}{\times} b := \overset{\gamma g}{\omega}(a) \underset{g}{} b - b \underset{g}{}\overset{\gamma g}{\omega}(a). \tag{4.51}$$

The field $\overset{\gamma g}{\omega}(a)$ is called the $\gamma\boldsymbol{g}$*-gauge rotation field.*

**Corollary 4.1** [2] Let $a \mapsto \boldsymbol{g}(a)$ be a metric extensor field defined on $U_o$, and let $a \mapsto \overset{\gamma g}{\omega}(a)$ a smooth $(1,2)$-extensor field defined on $U_o$. Define a smooth connection 2-extensor field on $U_o$, say $(a,b) \mapsto \gamma_a(b)$ by

$$\gamma_a(b) = \frac{1}{2}\boldsymbol{g}^{-1}(a \cdot \partial\boldsymbol{g})(b) + \overset{\gamma g}{\omega}(a) \underset{g}{\times} b, \tag{4.52}$$

for all smooth 1-form fields $a$ and $b$. Then, the triple $(U_o, \boldsymbol{g}, \gamma)$ is a Riemann-Cartan or Lorentz-Cartan *MCGSS* (depending on the signature of $\boldsymbol{g}$).

### 4.4.6 *Some Important Properties of a Metric Compatible Connection*

Given the *MCGSS* $(U_o, \boldsymbol{g}, \gamma)$ the operators $\nabla_a$ and $\nabla_a^-$ satisfy the following properties [1].

**(i)** *The Ricci Theorem* For all smooth 1-form fields $a, b, c$ we have

$$a \cdot \partial(\boldsymbol{g}(b) \cdot c) = \boldsymbol{g}(\nabla_a b) \cdot c + \boldsymbol{g}(b) \cdot \nabla_a c, \tag{4.53}$$

Ricci theorem may be generalized for arbitrary smooth multiform fields $X$ and $Y$, i.e.,

$$a \cdot \partial(\underline{g}(X) \cdot Y) = \underline{g}(\nabla_a X) \cdot Y + \underline{g}(X) \cdot \nabla_a Y. \tag{4.54}$$

**(ii)** *Leibniz rule f*or the metric scalar product (i.e., $X \underset{g}{\cdot} Y \equiv \underline{g}(X) \cdot Y$),

$$a \cdot \partial(X \underset{g}{\cdot} Y) = (\nabla_a^- X) \underset{g}{\cdot} Y + X \underset{g}{\cdot} (\nabla_a^- Y), \tag{4.55}$$

for all smooth multiform fields $X$ and $Y$.

**(iii)** For any smooth multiform field $X$,

$$\nabla_a^- X = \underline{g}(\nabla_a \underline{g}^{-1}(X)). \tag{4.56}$$

**(iv)** The action of the operators $\nabla_a$ and $\nabla_a^-$, on smooth extensor fields permits us to write the compatibility between the metric and the parallelism as

$$\nabla_a^- \boldsymbol{g} = 0, \tag{4.57}$$

$$\nabla_a \boldsymbol{g}^{-1} = 0. \tag{4.58}$$

### 4.4.7 *The Riemann 4-Extensor Field of a MCGSS* $(U_o, \boldsymbol{g}, \gamma)$

The smooth scalar 4-extensor field on $U_o$, $(a,b,c,w) \mapsto \overset{\gamma_g}{\mathbf{R}}_3(a,b,c,w)$, such that for all smooth 1-form fields $a, b, c$ and $w$,

$$\overset{\gamma_g}{\mathbf{R}}_3(a,b,c,w) = -\overset{\gamma_g}{\rho}(a,b,c) \cdot \boldsymbol{g}(w), \tag{4.59}$$

is called the *Riemann (curvature) 4-extensor field of the structure* $(U_o, \boldsymbol{g}, \gamma)$.

$\overset{\gamma_g}{\mathbf{R}}_3$ satisfies the following important properties [2]:

**(i)** On the parallelism structure $(U_o, \gamma)$, $\overset{\gamma_g}{\mathbf{R}}_3$ is antisymmetric with respect to the first and second variables, i.e.,

$$\overset{\gamma_g}{\mathbf{R}}_3(a,b,c,w) = -\overset{\gamma_g}{\mathbf{R}}_3(b,a,c,w). \tag{4.60}$$

**(ii)** On the parallelism structure $(U_o, \gamma)$, if the covariant derivative $\nabla_a$ is symmetric (i.e., $\nabla_a b - \nabla_b a = [a,b]$), then $\overset{\gamma_g}{\mathbf{R}}_3$ possesses a cyclic property, i.e.,

$$\overset{\gamma_g}{\mathbf{R}}_3(a,b,c,w) + \overset{\gamma_g}{\mathbf{R}}_3(b,c,a,w) + \overset{\gamma_g}{\mathbf{R}}_3(c,a,b,w) = 0. \tag{4.61}$$

**(iii)** On the *MCGSS* $(U_o, \boldsymbol{g}, \gamma)$, $\overset{\gamma_g}{\mathbf{R}}_3$ is antisymmetric with respect to the third and fourth variables, i.e.,

$$\overset{\gamma_g}{\mathbf{R}}_3(a,b,c,w) = -\overset{\gamma_g}{\mathbf{R}}_3(a,b,w,c). \tag{4.62}$$

**(iv)** On the *MCGSS* $(U_o, \boldsymbol{g}, \gamma)$, if the covariant derivative $\nabla_a$ is symmetric (i.e., $\nabla_a b - \nabla_b a = [a,b]$), then $\overset{\gamma_g}{\mathbf{R}}_3$

$$\overset{\gamma_g}{\mathbf{R}}_3(a,b,c,w) = \overset{\gamma_g}{\mathbf{R}}_3(c,w,a,b). \tag{4.63}$$

Note that, from the Eqs.(4.60) and (4.62), we have

$$\overset{\gamma_g}{\mathbf{R}}_3(a,b,c,w) = \overset{\gamma_g}{\mathbf{R}}_3(b,a,w,c) \tag{4.64}$$

and under those conditions, given a pair of reciprocal bases $(\{\varepsilon^\mu\}, \{\varepsilon_\mu\})$ on $U_o$, we agree in writing [3,4]

$$\overset{\gamma_g}{\mathbf{R}}_3(\varepsilon_\beta, \varepsilon^\alpha, \varepsilon_\rho, \varepsilon_\sigma) := R_{\beta\ \rho\sigma}^{\ \alpha}, \tag{4.65}$$

for the *components of the Riemann tensor*. Also, under the same conditions, Eq.(4.39) may be written as

$$\overset{\gamma_g}{\mathbf{R}}_2(a,b) = \overset{\gamma_g}{\mathbf{R}}_3(\partial_w, a, w, b) = \overset{\gamma_g}{\mathbf{R}}_3(a, \partial_w, b, w) \tag{4.66}$$

and we have

$$R_{\beta\rho} = \overset{\gamma_g}{\mathbf{R}}_2(\varepsilon_\beta, \varepsilon_\rho) = \overset{\gamma_g}{\mathbf{R}}_3(\varepsilon_\beta, \varepsilon^\alpha, \varepsilon_\rho, \varepsilon_\alpha) = R_{\beta\ \rho\alpha}^{\ \alpha} \tag{4.67}$$

for the *components of the Ricci tensor*.

The scalar field

$$\overset{\gamma_g}{R} = \overset{\gamma_g}{\mathbf{R}}_2(\boldsymbol{g}^{-1}(\partial_a), a), \tag{4.68}$$

i.e., a $\boldsymbol{g}^{-1}$-contraction between the first and second variables of $\overset{\gamma_g}{\mathbf{R}}_2$ is called the *Ricci scalar field* (or scalar curvature).

### *4.4.8 Existence Theorem for the on* $(U_o, \boldsymbol{g}, \gamma)$

**Proposition 4.1** [2] There exists a unique smooth (2, 2)-extensor field, say $B \mapsto \overset{\gamma g}{\mathbf{R}}_2(B)$, such that

$$\overset{\gamma g}{\mathbf{R}}_3(a,b,c,w) = \overset{\gamma g}{\mathbf{R}}_2(a \wedge b) \cdot (c \wedge w). \tag{4.69}$$

This field $B \mapsto \overset{\gamma g}{\mathbf{R}}_2(B)$ is called the Riemann (2, 2)-*extensor field* and we see that Eq.(4.69) may be thought of as a factorization of $\overset{\gamma g}{\mathbf{R}}_3$.

**Proposition 4.2** [2] The Ricci (1, 1)-extensor field $b \mapsto \overset{\gamma g}{\mathbf{R}}_1(b)$ and the Ricci scalar field $R$ may be written as $\boldsymbol{g}^{-1}$-divergences of the Riemann (2, 2)-extensor field $B \mapsto \overset{\gamma g}{\mathbf{R}}_2(B)$, i.e.,

$$\begin{aligned} \overset{\gamma g}{\mathbf{R}}_1(b) &= \boldsymbol{g}^{-1}(\partial_a) \lrcorner \overset{\gamma g}{\mathbf{R}}_2(a \wedge b), \\ \overset{\gamma g}{R} &= \boldsymbol{g}^{-1}(\partial_b) \cdot \overset{\gamma g}{\mathbf{R}}(b) = \underline{\boldsymbol{g}}^{-1}(\partial_a \wedge \partial_b) \cdot \overset{\gamma g}{\mathbf{R}}_2(a \wedge b). \end{aligned} \tag{4.70}$$

**Proof** A simple algebraic manipulation of Eq.(4.36) and Eq. (4.59) gives the extensor identity

$$\overset{\gamma g}{\mathbf{R}}_1(w,a,b,c) = \overset{\gamma g}{\mathbf{R}}_3(c,b,a,\boldsymbol{g}^{-1}(w)). \tag{4.71}$$

Now, according to Eq.(4.39) and Eq. (4.40), and taking also in account Eq.(4.71), we have

$$\begin{aligned} a \cdot \overset{g}{\mathbf{R}}_1(b) &= \overset{g}{\mathbf{R}}_2(a,b) \\ &= \overset{\gamma g}{\mathbf{R}}_1(\varepsilon^\mu, a, \varepsilon_\mu, b), \ (\mu \text{ summed from 0 to 3}) \\ &= \overset{\gamma g}{\mathbf{R}}_3(b, \varepsilon_\mu, a, \boldsymbol{g}^{-1}(\varepsilon^\mu)), \end{aligned} \tag{4.72}$$

and using essentially the factorization given by Eq.(4.69), we have

$$\begin{aligned} a \cdot \overset{\gamma g}{\mathbf{R}}_1(b) &= \overset{\gamma g}{\mathbf{R}}_2(b \wedge \varepsilon_\mu) \cdot (a \wedge \boldsymbol{g}^{-1}(\varepsilon^\mu)) = (\boldsymbol{g}^{-1}(\varepsilon^\mu) \wedge a) \cdot \overset{\gamma g}{\mathbf{R}}_2(\varepsilon_\mu \wedge b) \\ &= a \cdot (\boldsymbol{g}^{-1}(\varepsilon^\mu) \lrcorner \overset{g}{\mathbf{R}}_2(\varepsilon_\mu \wedge b)), \end{aligned}$$

i.e., $\overset{\gamma g}{\mathbf{R}}_1(b) = \boldsymbol{g}^{-1}(\varepsilon^\mu) \lrcorner \overset{\gamma g}{\mathbf{R}}_2(\varepsilon_\mu \wedge b) = \boldsymbol{g}^{-1}(\partial_a) \lrcorner \overset{\gamma g}{\mathbf{R}}_2(a \wedge b)$. ■

### *4.4.9 The Einstein (1, 1)-Extensor Field*

The smooth (1, 1)-extensor field, $a \mapsto \overset{\gamma g}{\mathbf{G}}(a)$, given by

$$\overset{\gamma g}{\mathbf{G}}(a) := \overset{\gamma g}{\mathbf{R}}_1(a) - \frac{1}{2} \boldsymbol{g}(a) \overset{\gamma g}{R}, \tag{4.73}$$

is called the *Einstein* (1, 1)-*extensor field.*

## 4.5 Riemann and Lorentz *MCGSS's* $(U_o, \boldsymbol{g}, \lambda)$

### *4.5.1 Levi-Civita Covariant Derivative*

It is important to have in mind that the unique *MCGSS* $(U_o, \boldsymbol{g}, \gamma)$, where the covariant derivative $\nabla_a$ is symmetric, is the one where the connection 2-extensor field $\gamma$ is precisely the Levi-Civita one, that we have already denoted by $\lambda$ (recall Eq.(4.47)). In this case, the *MCGSS* $(M, \mathbf{g}, \nabla = D)$ is said a Riemann or *Lorentz MCGSS* depending on the signature of $\mathbf{g}$. We also use this denomination for the structure $(U_o, \boldsymbol{g}, \lambda)$. The covariant derivatives defined by $\lambda$ are precisely the Levi-Civita covariant derivatives $D_a$ and $D_a^-$. Some properties of the curvature extensors of the *MCGSS* $(U_o, \boldsymbol{g}, \gamma)$, which are important for the purposes of this book, will be given below, and for that important case, *simplified* notations will be used.

The covariant derivative operators associated with the connection 2-extensor field $\lambda$ will be denoted by $D_a$ and $D_a^-$, and

$$D_a X := a \cdot \partial X + \Lambda_a(X), \tag{4.74a}$$

$$D_a^- X := a \cdot \partial X - \Lambda_a^\dagger(X), \tag{4.74b}$$

for all smooth multiform field $X \in \sec \bigwedge T^*U$, and called *Levi-Civita covariant derivative operators.* Note that $\Lambda_a$ is the *generalized* of $\lambda_a$ and $\Lambda_a^\dagger$ is the adjoint of $\Lambda_a$.

### *4.5.2 Properties of $D_a$*

**(i)** For all smooth 1-form fields $a, b$ and $c$ it is:

$$(D_a b) \cdot c = \begin{Bmatrix} c \\ a, b \end{Bmatrix}. \tag{4.75}$$

Indeed, using Eq.(4.74a) and Eq.(4.47), we have immediately

$$D_a b = a \cdot \partial b + \lambda_a(b) = a \cdot \partial b + \partial_c \begin{Bmatrix} c \\ a, b \end{Bmatrix} - a \cdot \partial b = \partial_c \begin{Bmatrix} c \\ a, b \end{Bmatrix}.$$

Now, using the formula for multiform differentiation $c \cdot \partial_n \phi(n) = c \cdot (\partial_n \phi(n))$, with $\phi$ a scalar function of 1-form variable, the formula
$c \cdot \partial_n f(n) = f(c)$, where $f$ is a multiform function of 1-form variable and taking into account the linearity of the Christoffel operator with respect to the superior argument, we have:

$$(D_a b) \cdot c = c \cdot (\partial_n \begin{Bmatrix} n \\ a, b \end{Bmatrix}) = c \cdot \partial_n \begin{Bmatrix} n \\ a, b \end{Bmatrix} = \begin{Bmatrix} c \\ a, b \end{Bmatrix}.$$

**(ii)** In order to become acquainted with the algebraic manipulations of the multiform and extensor calculus, we give the details in the derivation of the *Ricci theorem for the* Levi-Civita covariant derivative $D_a$, i.e.,

$$a \cdot \partial(\boldsymbol{g}(b) \cdot c) = \boldsymbol{g}(D_a b) \cdot c + \boldsymbol{g}(b) \cdot D_a c, \tag{4.76}$$

where $a, b, c \in \sec \bigwedge^1 T^*U$ are smooth 1-form fields.

Using Eq.(4.75), Eq.(4.42) and Eq.(4.46c), we have

$$\begin{aligned}\boldsymbol{g}(D_a b)\cdot c+\boldsymbol{g}(b)\cdot D_a c &= (D_a b)\cdot \boldsymbol{g}(c)+D_a c\cdot \boldsymbol{g}(b)\\ &= \begin{Bmatrix}\boldsymbol{g}(c)\\ a,b\end{Bmatrix}+\begin{Bmatrix}\boldsymbol{g}(b)\\ a,c\end{Bmatrix}\\ &= [a,b,\boldsymbol{g}^{-1}(\boldsymbol{g}(c))]+[a,c,\boldsymbol{g}^{-1}(\boldsymbol{g}(b))]\\ &= [a,b,c]+[a,c,b]=a\cdot\partial(\boldsymbol{g}(b)\cdot c).\end{aligned}$$

**(iii)** This theorem is also valid for smooth multiform fields $X, Y$, i.e.,

$$a\cdot\partial(X\cdot Y)=\underline{\boldsymbol{g}}(D_a X)\cdot Y+\underline{\boldsymbol{g}}(X)\cdot D_a Y, \tag{4.77}$$

**(iv)** The Levi-Civita covariant derivative $D_a$ is symmetric, i.e.,

$$D_a b - D_b a = [a,b], \tag{4.78}$$

for all smooth 1-form fields $a$ and $b$.

Take three smooth 1-form fields $a, b$ and $c$. Using Eq.(4.75), Eq.(4.44) and Eq.(4.46b), we can write

$$\begin{aligned}(D_a b)\cdot c-(D_b a)\cdot c &= \begin{Bmatrix}c\\ a,b\end{Bmatrix}-\begin{Bmatrix}c\\ b,a\end{Bmatrix}=[a,b,\boldsymbol{g}^{-1}(c)]-[b,a,\boldsymbol{g}^{-1}(c)]\\ &= \boldsymbol{g}(\boldsymbol{g}^{-1}(c))\cdot[a,b]=[a,b]\cdot c,\end{aligned}$$

which from the non degeneracy of the scalar product gives

$$D_a b - D_b a = [a,b]. \blacksquare$$

**(v)** The Levi-Civita covariant derivative $D_a$ is $\boldsymbol{g}$-*compatible,* i.e.,

$$D_a^- \boldsymbol{g}=0. \tag{4.79}$$

Take three smooth 1-form fields $a, b$ and $c$. Using Eq.(4.29) and the Ricci theorem (Eq.(4.76)), we have

$$\begin{aligned}(D_a^- \boldsymbol{g})(b)\cdot c &= a\cdot\partial(\boldsymbol{g}(b)\cdot c)-\boldsymbol{g}(D_a b)\cdot c-\boldsymbol{g}(b)\cdot D_a c,\\ &= 0,\end{aligned}$$

which implies that $(D_a^- \boldsymbol{g})(b)=0$, i.e., $D_a^- \boldsymbol{g}=0$.

We emphasize that the symmetry property and the $\boldsymbol{g}$-compatibility uniquely characterizes the Levi-Civita covariant derivative, i.e., there exists a unique pair of covariant derivative operators $\nabla_a$ and $\nabla_a^-$ such that $\nabla_a$ is symmetric (i.e., $\nabla_a b-\nabla_b a=[a,b]$) and $\nabla_a^- \boldsymbol{g}=0$ (i.e., $\nabla_a^-$ is $\boldsymbol{g}$-compatible). Those $\nabla_a$ and $\nabla_a^-$ are precisely $D_a$ and $D_a^-$.

### 4.5.3 Properties of $\boldsymbol{R}_2(B)$ and $\boldsymbol{R}_1(b)$

On a Riemann or Lorentz *MCGSS* $(U_o, \boldsymbol{g}, \lambda)$, we use simplified notations, the Riemann $(2,2)$-extensor field is denoted $B \mapsto \boldsymbol{R}_2(B)$ and the Ricci $(1,1)$-extensor is denoted $b \mapsto \boldsymbol{R}_1(b)$. These objects now have three additional properties besides the ones those objects have on a general *MCGSS* $(U_o, \boldsymbol{g}, \gamma)$. We have

**(i)** $\boldsymbol{R}_2(B)$ is adjoint symmetric, i.e.,

$$\boldsymbol{R}_2(B) = \boldsymbol{R}_2^{\dagger}(B). \tag{4.80}$$

**(ii)** $\boldsymbol{R}_1(b)$ is adjoint symmetric, i.e.,

$$\boldsymbol{R}_1(b) = \boldsymbol{R}_1^{\dagger}(b). \tag{4.81}$$

**(iii)** The *matrix element* for the Ricci extensor field is given by the notable formula

$$\boldsymbol{R}_1(b) \cdot c = \partial_a \cdot \rho(a, b, c), \tag{4.82}$$

i.e., the divergent of the curvature operator with respect to the first variable.

### 4.5.4 Levi-Civita Differential Operators

On the Riemann or Lorentz *GSS* $(U_o, \boldsymbol{g}, \lambda)$, we can introduce three differential operators : the gradient $\underset{g}{\boldsymbol{D}}$, the divergent $\underset{g}{\boldsymbol{D}}\lrcorner$ and the *rotational* $\boldsymbol{D}\wedge$, acting on smooth multiform fields.

- The *gradient of a smooth multiform field* $X \in \sec \bigwedge T^*U$ is defined by

$$\underset{g}{\boldsymbol{D}} X := \partial_a \underset{g}{(} D_a^- X), \tag{4.83}$$

  i.e., $\underset{g}{\boldsymbol{D}} X = \varepsilon^{\mu} \underset{g}{(} D_{\varepsilon_\mu}^- X) = \varepsilon_{\mu} \underset{g}{(} D_{\varepsilon^\mu}^- X)$, where $(\{\varepsilon^\mu\}, \{\varepsilon_\mu\})$ is a pair of arbitrary reciprocal bases defined on $U_o$.

- The *divergent of a smooth multiform field* $X \in \sec \bigwedge T^*U$ is defined by

$$\underset{g}{\boldsymbol{D}}\lrcorner X = \partial_a \underset{g}{\lrcorner} (D_a^- X). \tag{4.84}$$

- The *rotational of a smooth multiform field* $X \in \sec \bigwedge T^*U$ is defined by

$$\boldsymbol{D} \wedge X = \partial_a \wedge (D_a^- X). \tag{4.85}$$

The main properties of those operators are:

**(i)** For any smooth multiform field $X \in \sec \bigwedge T^*U$,

$$\underset{g}{\boldsymbol{D}} X = \underset{g}{\boldsymbol{D}}\lrcorner X + \boldsymbol{D} \wedge X, \tag{4.86}$$

i.e., $\underset{g}{\boldsymbol{D}} = \underset{g}{\boldsymbol{D}}\lrcorner + \boldsymbol{D}\wedge$.

**(ii)** For any smooth multiform field $X \in \sec \bigwedge T^*U$

$$\underset{g}{\boldsymbol{D}}\lrcorner X = \frac{1}{\sqrt{|\det[\boldsymbol{g}]|}} \underline{g}(\partial \lrcorner \sqrt{|\det[\boldsymbol{g}]|}\, \underline{g}^{-1}(X)), \tag{4.87}$$

$$\boldsymbol{D} \wedge X = \partial \wedge X. \tag{4.88}$$

## 4.6 Deformation of *MCGSS* Structures

### 4.6.1 Enter the Plastic Distortion Field $\mathbf{h}$

Minkowski Metric on $U_o$

Let $M$ be a smooth manifold with $\dim M = 4$. Consider the canonical spaces $\mathbf{U}$ and $U$ defined by local coordinates $(\mathcal{U}, \phi)_o$, and the canonical dual bases[6] $\{\mathbf{b}_\mu\}$ and $\{\boldsymbol{\beta}^\mu\}$ for $\mathbf{U}$ and $U$, $\boldsymbol{\beta}^\mu(\mathbf{b}_\nu) = \delta^\mu_\nu$. Moreover, we denote the reciprocal basis of the basis $\{\boldsymbol{\beta}^\mu\}$ by $\{\boldsymbol{\beta}_\mu\}$, with $\boldsymbol{\beta}^\mu \cdot \boldsymbol{\beta}_\nu = \delta^\mu_\nu$. Note that due to the identifications that define the canonical space, we can also write that $\boldsymbol{\beta}^\mu, \boldsymbol{\beta}_\nu \in \sec \bigwedge^1 T^*U$ for all $x \in U_0$.

Recalling Section 2.8.4 we will introduce the smooth extensor field on $U_0$, say $\eta \in \sec(1,1)\text{-}extU$, defined for any $a \in \sec \bigwedge^1 T^*U$ by $a \mapsto \eta(a) \in \sec \bigwedge^1 T^*U$, such that

$$\eta(a) = \boldsymbol{\beta}^0 a \boldsymbol{\beta}^0. \tag{4.89}$$

The field $\eta$ is a well defined metric extensor field on $U_0$ (i.e., $\eta$ is symmetric and non degenerate) and it is called the *Minkowski metric extensor*[7].

The main properties of $\eta$ are:

**(i)** $\eta$ is a constant $(1,1)$-extensor field, i.e., $a \cdot \partial \eta = 0$.

**(ii)** The 1-forms of the canonical basis $\{\boldsymbol{\beta}^\mu\}$ are $\eta$ orthonormal (i.e., $\eta(\boldsymbol{\beta}^\mu) \cdot \boldsymbol{\beta}^\nu = \delta^{\mu\nu}$); $\boldsymbol{\beta}^0$ is an eigenvector with eigenvalue 1, i.e., $\eta(\boldsymbol{\beta}^0) = \boldsymbol{\beta}^0$, and $\boldsymbol{\beta}^k$ $(k = 1,2,3)$ are eigenvectors with eigenvalues $-1$, i.e., $\eta(\boldsymbol{\beta}^k) = -\boldsymbol{\beta}^k$. Then, $\eta$ has signature $(1,3)$.

**(iii)** $\operatorname{tr}[\eta] = -2$ and $\det[\eta] = -1$.

**(iv)** The extended of $\eta$ has the same generator of $\eta$, i.e., for any $X \in \sec \bigwedge^1 T^*U$, $\underline{\eta}(X) = \boldsymbol{\beta}^0 X \boldsymbol{\beta}^0$.

**(v)** $\eta$ is orthogonal canonical, i.e., $\eta = \eta^\clubsuit$ (recall that $\eta^\clubsuit = (\eta^\dagger)^{-1} = (\eta^{-1})^\dagger$). Then, $\eta^2 = i_d$.

Lorentzian Metric

Given a metric tensor $\mathbf{g} \in \sec T^0_2 M$ of signature $(1,3)$, if the $(1,1)$-extensor field associated with $\mathbf{g}$ is $\mathbf{g}$ and its representative on $U_0$ is $\boldsymbol{g}$, then $\boldsymbol{g}$ also has the same signature as the Minkowski metric extensor, i.e., signature$(\boldsymbol{g})$ = signature$(\eta) = (1,3)$. In what follows, $\mathbf{g}$ (and its representative $\boldsymbol{g}$) will be called a *Lorentzian metric extensor field*, or simply Lorentzian metric, if the context is clear.

Recalling Section 2.8.4, we know that there exists a (non unique) $(1,1)$-extensor field $\boldsymbol{h}$ given by:

$$\boldsymbol{h}(a) = \sum_{\mu=0}^{3} \sqrt{|\lambda^\mu|}(a \cdot v^\mu)\boldsymbol{\beta}^\mu, \tag{4.90}$$

where the $\lambda^\mu \in \sec \bigwedge^0 T^*U$ and $v^\mu \in \sec \bigwedge^1 T^*U$ are respectively the eigenvalue fields and the associated eigenvector fields of $\boldsymbol{g}$, such that

$$\boldsymbol{g} = \boldsymbol{h}^\dagger \eta \boldsymbol{h}. \tag{4.91}$$

[6]For spacetime structures, as usual the Greek indices take the values $0,1,2,3$ and Latin indices take the values $1,2,3$.

[7]We eventually call $\eta$ the Minkowski metric, when the context is clear.

The set of 1-forms $\{v^\mu\}$ define a *Euclidean orthonormal basis* for $V$, i.e., $v^\mu \cdot v^\nu = \delta^{\mu\nu}$. As we know, $\boldsymbol{h}$ is not unique and is defined modulo a local Lorentz transformation, i.e., $\boldsymbol{h}$ and $\boldsymbol{h}' = \Lambda \boldsymbol{h}$ determines the same $\boldsymbol{g}$, if $\Lambda \in \sec(1,1)\text{-}extU$ is a local Lorentz transformation, i.e., for any $a, b \in \sec \bigwedge^1 T^*U$, $\Lambda(a) \underset{\eta}{\cdot} \Lambda(b) = a \underset{\eta}{\cdot} b$. In what follows, we call $\boldsymbol{h}$ a *plastic gauge distortion field* on $U_o$.

## *4.6.2 On Elastic and Plastic Deformations*

The wording gauge is, of course, well justified, since $\boldsymbol{h}$ can only be determined modulus a local Lorentz transformation. The wording distortion is also well justified in view of Eq.(4.91), $\boldsymbol{h}$ distorts $\eta$ into $\boldsymbol{g}$. The wording plastic is justified as follows. In the theory of deformations and defects on continuum media [5], two different kinds of deformations are introduced for any medium that exists on a manifold $M$ carrying a metric field, say $\mathring{\mathbf{g}}$. These deformations are: (i) an *elastic* one, where the deformation of the medium is described by a diffeomorphism $\mathtt{h}$: $M \to M$ which induces a *deformed* metric given by the pullback metric[8] $\mathbf{g} = \mathtt{h}^* \, \mathring{\mathbf{g}}$ on $M$ and which is used in the definition of the *Cauchy-Green tensor* [6], and (ii) a *plastic* one, where the deformation of the medium is such that the new effective metric $\mathbf{g}$ on $M$ defining distance measurements *cannot* be described by the pullback of any diffeomorphism, as in the elastic case. In case (i) we can show that the pullback $D' = \mathtt{h}^* D$ of the Levi-Civita connection $D$ of $\mathring{\mathbf{g}}$, has the same torsion and Riemann curvature tensors as $D$, i.e., they are null if $D$ has null curvature and torsion tensors[9], whereas in case (ii) the medium can conveniently be described by a new effective *MCGSS* with a connection $\nabla$ that is $\boldsymbol{g}$ compatible ($\nabla \boldsymbol{g} = 0$), but that has in general non null torsion and Riemann curvature tensors.

Now, although there exists a distortion field $\boldsymbol{h}$ corresponding to the pullback $\mathtt{h}^*$ of any diffeomorphism [2], there are distortion fields $\boldsymbol{h}$ to which there corresponds *no* diffeomorphism. Thus, the wording *plastic* is justified, even more because we are going to see below that $\boldsymbol{h}$ also deforms in a precise *sense* the Levi-Civita connection of $\mathring{\mathbf{g}}$ and thus produces a connection on $M$ that is not compatible with $\mathring{\mathbf{g}}$. And, in this sense we can also say that $\boldsymbol{h}$ distorts the parallelism structure defined by the Levi-Civita connection of $\mathring{\mathbf{g}}$.

Construction of a Lorentzian Metric Field on $U_o$

**Proposition 4.1** [1] Let $\boldsymbol{h}$ be a $(1,1)$-extensor field on $U_0$, defined by

$$\boldsymbol{h}(a) = \sum_{\mu=0}^{3} \rho^\mu \Lambda(a) \cdot \beta^\mu \beta^\mu, \tag{4.92}$$

where $\rho^\mu \in \sec \bigwedge^0 T^*U$ are positive scalar fields on $U_o$ (i.e., $\rho^\mu(x) > 0$, for $x \in U_0$) and $\Lambda$ is an $\eta$-orthogonal $(1,1)$-extensor field on $U_0$ (i.e., $\Lambda(v) \underset{\eta}{\cdot} \Lambda(w) = \Lambda(v) \underset{\eta}{\cdot} \Lambda(w)$, for all vector fields $v$ and $w$); then, the $(1,1)$-extensor field $\boldsymbol{g}$ on $U_o$ given by

$$\boldsymbol{g} = \boldsymbol{h}^\dagger \eta \boldsymbol{h}, \tag{4.93}$$

[8] Take notice that here $\mathtt{h}^*$ denotes the pulback mapping.

[9] This result, which can be easily proven is shown, e.g, in Remark 250 in [4]. We take the opportunity to call the reader's attention that in [7] it is presented a theory for the gravitational field where the field $\mathbf{g}$ has been identified as a result of an elastic deformation. Such identification is, of course, incorrect. The arXiv version of the paper [http://arxiv.org/abs/math-ph/0608017] contains a correction.

is a Lorentzian metric extensor field (i.e., signature($\boldsymbol{g}$) = (1,3)).

The non null scalar fields $(\rho^0)^2$ and $-(\rho^1)^2, -(\rho^2)^2, -(\rho^3)^2$ are the eigenvalue fields of $\boldsymbol{g}$, and the non null 1-form fields $\Lambda^\dagger(\boldsymbol{\beta}^0)$ and $\Lambda^\dagger(\boldsymbol{\beta}^1), \boldsymbol{\beta}^\dagger(\beta^2)$, $\boldsymbol{\beta}^\dagger(\beta^3)$ are the associated eigenvector fields of $\boldsymbol{g}$.

## 4.7 Deformation of a Minkowski-Cartan *MCGSS* into a Lorentz-Cartan *MCGSS*

Let $M$ be 4-dimensional (with enough structure to be part of a Lorentzian spacetime structure as defined in Section 1). Let $(\mathcal{U}, \phi)_o$ a chart for $\mathcal{U} \subset M$ containing $O \in \mathcal{U}$ and let $U_o \subset U$ be defined as previously. A *MCGSS* $(U_o, \eta, \varkappa)$, where $\eta$ is the Minkowski metric extensor on $U_o$ and $\varkappa$ is a connection extensor field on $U_o$ compatible with $\eta$, is said to be a *Minkowski-Cartan MCGSS.*

The metric compatibility may be written $\overset{\varkappa\eta^-}{\mathcal{D}}_a \eta = 0$, where $\overset{\varkappa\eta^-}{\mathcal{D}}_a$ is one of the covariant derivative operators associated with $\varkappa$. Before going on, a crucial observation is needed.

**Remark 4.3** Note that we did not impose the condition that the manifold $M$ carries a Minkowski metric field $\boldsymbol{\eta}$, as *given* by Eq.(1.6). If that were the case, we would have $M \simeq \mathbb{R}^4$, which is not being considered at this moment.

Now, let us recall that the Theorem 4.1 (Eq.(4.50)) implies the existence of a *$\varkappa\eta$-gauge rotation field*, say $\overset{\varkappa\eta}{\Omega}$ such

$$\varkappa_a(b) = \overset{\varkappa\eta}{\Omega}(a) \underset{\eta}{\times} b = \overset{\varkappa\eta}{\Omega}(a) \times \eta(b). \tag{4.94}$$

Also, according to Eq.(4.12a), the covariant derivative $\overset{\varkappa\eta}{\mathcal{D}}_a$ when acting on a smooth 1-form field is given by

$$\overset{\varkappa\eta}{\mathcal{D}}_a b = a \cdot \partial b + \overset{\varkappa\eta}{\Omega}(a) \underset{\eta}{\times} b. \tag{4.95}$$

**Remark 4.4** Keep in mind that, in a general Minkowski-Cartan *GSS* $(U_o, \eta, \varkappa)$, $\overset{\varkappa\eta}{\mathcal{D}}_a$ is not the representative of the Levi-Civita connection of $\boldsymbol{\eta}$ and indeed, in general, possesses non null torsion and Riemann curvature tensors.

We recall that a *MCGSS* $(M, \mathbf{g}, \nabla)$, where $M$ is a 4-dimensional manifold as in the previous subsection, $\mathbf{g}$ is a Lorentzian metric, $\nabla$ is an arbitrary metric compatible connection on $M$, i.e., $\nabla \mathbf{g} = 0$. It is said to be a *Lorentz-Cartan MCGSS.* Let $U_o$, $\boldsymbol{g}$, and $\gamma$ be as defined previously. We recall that we agreed in calling also $(U_o, \boldsymbol{g}, \gamma)$ a Lorentz-Cartan *MCGSS*.

The compatibility between $\boldsymbol{g}$ and $\gamma$ may be written as $\overset{\gamma g}{\nabla}{}_a^- \boldsymbol{g} = 0$, where $\overset{\gamma g}{\nabla}{}_a^-$ is one of the covariant derivative operators associated to $\gamma$.

The Theorem 4.1 implies in this case the existence of a $\gamma\boldsymbol{g}$-gauge rotation field $\overset{\gamma g}{\omega}(a)$, such that

$$\gamma_a(b) = \frac{1}{2}\boldsymbol{g}^{-1}(a \cdot \partial \boldsymbol{g})(b) + \overset{\gamma g}{\omega}(a) \underset{g}{\times} b = \frac{1}{2}\boldsymbol{g}^{-1}(a \cdot \partial \boldsymbol{g})(b) + \overset{\gamma g}{\omega}(a) \times \boldsymbol{g}(b). \tag{4.96}$$

According to Eq.(4.12a), the covariant derivative $\overset{\gamma g}{\nabla}_a$, when acting on a smooth 1-form field, is given by

$$\overset{\gamma g}{\nabla}_a b = a \cdot \partial b + \frac{1}{2}\boldsymbol{g}^{-1}(a \cdot \partial \boldsymbol{g})(b) + \overset{\gamma g}{\omega}(a) \underset{g}{\times} b. \tag{4.97}$$

### *4.7.1* ***h**-Distortions of Covariant Derivatives*

**Theorem 4.2**. The operators $\overset{\gamma g}{\nabla}_a$ and $\overset{\gamma g}{\nabla}{}^{-}_{a}$ of the Lorentz-Cartan *GSS* are related with the operators $\overset{\varkappa\eta}{\mathcal{D}}_a$ and $\overset{\varkappa\eta-}{\mathcal{D}}_a$ of the Minkowski-Cartan *GSS* by

$$\overset{\gamma g}{\nabla}_a b = \boldsymbol{h}^{-1}(\overset{\varkappa\eta}{\mathcal{D}}_a \boldsymbol{h}(b)), \tag{4.98}$$

$$\overset{\gamma g}{\nabla}{}^{-}_{a} b = \boldsymbol{h}^{\dagger}(\overset{\varkappa\eta-}{\mathcal{D}}_a \boldsymbol{h}^{\clubsuit}(b)), \tag{4.99}$$

where $\boldsymbol{h}$ is the *plastic distortion field* introduced above such that $\boldsymbol{g} = \boldsymbol{h}^{\dagger}\eta\boldsymbol{h}$.

In order to prove the Theorem 4.2 we shall need the results of the following two lemmas.

**Lemma 4.1** Let $(\overset{1}{\nabla}_a, \overset{1}{\nabla}{}^{-}_{a})$ be any pair of associated covariant derivative operators, i.e., $\overset{1}{\nabla}_a$ and $\overset{1}{\nabla}{}^{-}_{a}$ satisfy Eq.(4.18),

$$a \cdot \partial(b \cdot c) = (\overset{1}{\nabla}_a b) \cdot c + b \cdot \overset{1}{\nabla}{}^{-}_{a} c. \tag{4.100}$$

Then, the pair of covariant derivative operators $(\overset{2}{\nabla}, \overset{2}{\nabla}{}^{-}_{a})$, defined by

$$\overset{2}{\nabla}_a b = \boldsymbol{h}^{-1}(\overset{1}{\nabla}_a \boldsymbol{h}(b)), \tag{4.101}$$

$$\overset{2}{\nabla}{}^{-}_{a} b = \boldsymbol{h}^{\dagger}(\overset{1}{\nabla}{}^{-}_{a} \boldsymbol{h}^{\clubsuit}(b)), \tag{4.102}$$

where $\boldsymbol{h}$ is a smooth invertible $(1,1)$-extensor field is also a pair of associated covariant derivative operators, i.e., $\overset{2}{\nabla}_a$ and $\overset{2}{\nabla}{}^{-}_{a}$ satisfy Eq.(4.18),

$$a \cdot \partial(b \cdot c) = (\overset{2}{\nabla}_a b) \cdot c + b \cdot (\overset{2}{\nabla}{}^{-}_{a} c). \tag{4.103}$$

**Proof** Given that $\overset{1}{\nabla}_a$ and $\overset{1}{\nabla}{}^{-}_{a}$ are well defined covariant derivative operators they satisfy Eqs.(4.13), (4.14) and (4.15). Then, a simple algebraic manipulation shows that $\overset{2}{\nabla}_a$ and $\overset{2}{\nabla}{}^{-}_{a}$ also satisfy those same equations. Thus, $\overset{2}{\nabla}_a$ and $\overset{2}{\nabla}{}^{-}_{a}$ are also well defined covariant derivative operators.

On the other side, using Eq.(4.18), we have

$$\overset{1}{\nabla}_a \boldsymbol{h}(b) \cdot \boldsymbol{h}^{\clubsuit}(c) + \boldsymbol{h}(b) \cdot \overset{1}{\nabla}{}^{-}_{a} \boldsymbol{h}^{\clubsuit}(c) = a \cdot \partial(\boldsymbol{h}(b) \cdot \boldsymbol{h}^{\clubsuit}(c)),$$

and after some algebraic manipulations, using Eqs.(4.101) and (4.102), we get

$$\boldsymbol{h}^{-1}(\overset{1}{\nabla}_a \boldsymbol{h}(b)) \cdot c + b \cdot \boldsymbol{h}^{\dagger}(\overset{1}{\nabla}{}^{-}_{a} \boldsymbol{h}^{\clubsuit}(c)) = a \cdot \partial(\boldsymbol{h}^{-1}\boldsymbol{h}(b) \cdot c)$$
$$(\overset{2}{\nabla}_a b) \cdot c + b \cdot \overset{2}{\nabla}{}^{-}_{a} c = a \cdot \partial(b \cdot c).$$

We see then that $\overset{2}{\nabla}_a$ and $\overset{2}{\nabla}{}^{-}_{a}$ also satisfy Eq.(4.18), i.e., $(\overset{2}{\nabla}_a, \overset{2}{\nabla}{}^{-}_{a})$ is a pair of associated covariant derivative operators. ∎

**Lemma 4.2** Let $\boldsymbol{h}$ be a plastic distortion field such that $\boldsymbol{g} = \boldsymbol{h}^{\dagger}\eta\boldsymbol{h}$, and let $\overset{1}{\nabla}{}_a^-$ and $\overset{2}{\nabla}{}_a^-$ be a pair of associated covariant derivative operators acting on the module of the smooth $(p,q)$-extensor fields. Then

$$\overset{2}{\nabla}{}_a^- \boldsymbol{g} = \boldsymbol{h}^{\dagger}(\overset{1}{\nabla}{}_a^- \eta)\boldsymbol{h}. \tag{4.104}$$

**Proof** Indeed, according to the Eq.(4.19b),utilizing Eqs.(4.101) and (4.102) and also taking into account the Lemma 4.1, we have

$$\begin{aligned}(\overset{2}{\nabla}{}_a^- \boldsymbol{g})(b) &= \overset{2}{\nabla}{}_a^- \boldsymbol{g}(b) - \boldsymbol{g}(\overset{2}{\nabla}_a b) = \boldsymbol{h}^{\dagger}\overset{1}{\nabla}{}_a^- \boldsymbol{h}^{\clubsuit}\boldsymbol{h}^{\dagger}\eta\boldsymbol{h}(b) - \boldsymbol{h}^{\dagger}\eta\boldsymbol{h}\boldsymbol{h}^{-1}\overset{1}{\nabla}_a\boldsymbol{h}(b)\\ &= \boldsymbol{h}^{\dagger}(\overset{1}{\nabla}{}_a^- \eta\boldsymbol{h}(b)) - \eta(\overset{1}{\nabla}_a\boldsymbol{h}(b)) = \boldsymbol{h}^{\dagger}(\overset{1}{\nabla}{}_a^- \eta)\boldsymbol{h}(b),\end{aligned}$$

i.e., $\overset{2}{\nabla}{}_a^- \boldsymbol{g} = \boldsymbol{h}^{\dagger}(\overset{1}{\nabla}{}_a^- \eta)\boldsymbol{h}$. ∎

Eq.(4.104) shows that if $\overset{1}{\nabla}{}_a^-$ is $\eta$-compatible (i.e., $\overset{1}{\nabla}{}_a^- \eta = 0$) iff $\overset{2}{\nabla}{}_a^-$ is $\boldsymbol{g}$-compatible (i.e., $\overset{2}{\nabla}{}_a^- \boldsymbol{g} = 0$). Or, in other words, the operators $\overset{1}{\nabla}_a$ and $\overset{1}{\nabla}{}_a^-$ are covariant derivative operators on the Minkowski-Cartan *MCGSS* iff $\overset{2}{\nabla}_a$ and $\overset{2}{\nabla}{}_a^-$ are covariant derivative operators on the Lorentz-Cartan *MCGSS*.

**Proof of Theorem 4.2** The theorem follows at once from Lemmas 4.1. and 4.2 using the identifications:

$$\overset{\varkappa\eta}{\mathcal{D}}_a \equiv \overset{1}{\nabla}_a,\ \overset{\varkappa\eta^-}{\mathcal{D}}_a \equiv \overset{1}{\nabla}{}_a^- \text{ and } \overset{\gamma_g}{\nabla}_a \equiv \overset{2}{\nabla}_a, \overset{\gamma_g}{\nabla}{}_a^- \equiv \overset{2}{\nabla}{}_a^-. \ ∎$$

## 4.8 Coupling Between the Minkowski-Cartan and the Lorentz-Cartan *MCGSS*

**Proposition 4.3** There exists a smooth 2-extensor field on $U_0$, such that for all $a \in \sec \bigwedge^1 T^*U$, there exists $f_a : \sec \bigwedge^1 T^*U \to \sec \bigwedge^1 T^*U$, $b \mapsto f_a(b)$, which is $\eta$-adjoint antisymmetric [10] (i.e., $f_a = -f_a^{\dagger(\eta)}$), such that

$$\underline{\boldsymbol{h}}\overset{\gamma_g}{\omega}(a) = \overset{\varkappa\eta}{\Omega}(a) + \frac{1}{2}\overset{\eta}{\text{bif}}[f_a]. \tag{4.105}$$

Such a 2-extensor field is given by the formula

$$f_a(b) = (a \cdot \partial\boldsymbol{h})\boldsymbol{h}^{-1}(b) - \frac{1}{2}\eta\boldsymbol{h}^{\clubsuit}(a \cdot \partial\boldsymbol{g})\boldsymbol{h}^{-1}(b). \tag{4.106}$$

The proof is a simple calculation.

---

[10] The $\boldsymbol{g}$-adjoint (or metric adjoint) of a $(1,1)$-extensor $t$ is the $(1,1)$-extensor $t^{\dagger(\boldsymbol{g})} \equiv \boldsymbol{g} \circ t^{\dagger} \circ \boldsymbol{g}^{-1}$.

### 4.8.1 The Gauge Riemann and Ricci Fields

Let $(a,b,c) \mapsto \overset{\varkappa\eta}{\rho}(a,b,c)$ and $(a,b,c) \mapsto \overset{\gamma g}{\rho}(a,b,c)$ be the curvature operators of the *MCGSS*'s $(U_o, \eta, \varkappa)$ and $(U_o, \boldsymbol{g}, \gamma)$ as previously defined and $(a,b,c,w) \mapsto \overset{\varkappa\eta}{\mathbf{R}}_3(a,b,c,w)$ and $(a,b,c,w) \mapsto \overset{\gamma g}{\mathbf{R}}_3(a,b,c,w)$ the corresponding Riemann 4-extensor fields.

Let $B \mapsto \overset{\varkappa\eta}{\mathbf{R}}_2(B)$ and $B \mapsto \overset{\gamma g}{\mathbf{R}}_2(B)$ be the Riemann 2-extensor fields of $(M, \eta, \varkappa)$ and $(M, \boldsymbol{g}, \gamma)$.

We related $\overset{\varkappa\eta}{\rho}(a,b,c)$ with $\overset{\gamma g}{\rho}(a,b,c)$, using Eq.(4.98). Indeed,

$$\begin{aligned}
\overset{\gamma g}{\rho}(a,b,c) &= [\nabla_a, \nabla_b]c - \nabla_{[a,b]}c \\
&= \boldsymbol{h}^{-1}([\mathcal{D}_a, \mathcal{D}_b]\boldsymbol{h}(c) - \mathcal{D}_{[a,b]}\boldsymbol{h}(c)) \\
\overset{\gamma g}{\rho}(a,b,c) &= \boldsymbol{h}^{-1}(\overset{\varkappa}{\rho}(a,b,\boldsymbol{h}(c))).
\end{aligned} \tag{4.107}$$

We relate $\overset{\varkappa\eta}{\mathbf{R}}_3(a,b,c,w)$ with $\overset{\gamma g}{\mathbf{R}}_3(a,b,c,w)$, using Eq.(4.107). Indeed,

$$\begin{aligned}
\overset{\gamma g}{\mathbf{R}}_3(a,b,c,w) &= -\overset{\gamma g}{\rho}(a,b,c) \cdot \boldsymbol{g}(w) \\
&= -\boldsymbol{h}^{-1}(\overset{\varkappa\eta}{\rho}(a,b,\boldsymbol{h}(c))) \cdot \boldsymbol{h}^\dagger \eta \boldsymbol{h}(w) \\
&= -\overset{\varkappa\eta}{\rho}(a,b,\boldsymbol{h}(c)) \cdot \eta \boldsymbol{h}(w) \\
\overset{\gamma g}{\mathbf{R}}_3(a,b,c,w) &= \overset{\varkappa\eta}{\mathbf{R}}_3(a,b,\boldsymbol{h}(c),\boldsymbol{h}(w)).
\end{aligned} \tag{4.108}$$

We related $\overset{\varkappa\eta}{\boldsymbol{R}}_2(B)$ with $\overset{\gamma g}{\boldsymbol{R}}_2(B)$, using the factorization of $\mathbf{R}_3$, and the Eqs.(4.69) and (4.108). Indeed,

$$\begin{aligned}
&\overset{\gamma g}{\boldsymbol{R}}_2(a \wedge b) \cdot (c \wedge w) \\
&= \overset{\gamma g}{\mathbf{R}}_3(a,b,c,w) \\
&= \overset{\varkappa\eta}{\mathbf{R}}_3(a,b,\boldsymbol{h}(c),\boldsymbol{h}(w)) \\
&= \overset{\varkappa\eta}{\boldsymbol{R}}_2(a \wedge b) \cdot (\boldsymbol{h}(c) \wedge \boldsymbol{h}(w)) \\
&= \underline{h}^\dagger \overset{\varkappa\eta}{\boldsymbol{R}}_2(a \wedge b) \cdot (c \wedge w),
\end{aligned}$$

which implies

$$\overset{\gamma g}{\boldsymbol{R}}_2(a \wedge b) = \underline{\boldsymbol{h}}^\dagger \overset{\varkappa\eta}{\boldsymbol{R}}_2(a \wedge b),$$

i.e.,

$$\overset{\gamma g}{\boldsymbol{R}}(B) = \underline{\boldsymbol{h}}^\dagger \overset{\varkappa\eta}{\boldsymbol{R}}_2(B). \tag{4.109}$$

Now we get $\overset{\varkappa\eta}{\rho}(a,b,c)$ and $\overset{\varkappa\eta}{\boldsymbol{R}_2}(a\wedge b)$ in terms of the $\varkappa\eta$-gauge rotation field $\overset{\varkappa\eta}{\Omega}$ of the Minkowski-Cartan structure. For $\overset{\varkappa\eta}{\rho}(a,b,c)$, we have (recalling Eq.(4.95)):

$$\begin{aligned}\overset{\varkappa\eta}{\rho}(a,b,c) &= [\overset{\varkappa}{\mathcal{D}}_a, \overset{\varkappa}{\mathcal{D}}_b]c - \overset{\varkappa}{\mathcal{D}}_{[a,b]}c \\ &= (a\cdot\partial\overset{\varkappa\eta}{\Omega}(b) - b\cdot\partial\overset{\varkappa\eta}{\Omega}(a) - \overset{\varkappa\eta}{\Omega}([a,b]) + \overset{\varkappa\eta}{\Omega}(a)\underset{\eta}{\times}\overset{\varkappa\eta}{\Omega}(b))\underset{\eta}{\times} c\end{aligned}$$

$$\overset{\varkappa\eta}{\rho}(a,b,c) = ((a\cdot\partial\overset{\varkappa\eta}{\Omega})(b) - (b\cdot\partial\overset{\varkappa\eta}{\Omega})(a) + \overset{\varkappa\eta}{\Omega}(a)\underset{\eta}{\times}\overset{\varkappa\eta}{\Omega}(b))\underset{\eta}{\times} c. \tag{4.110}$$

For $\overset{\varkappa\eta}{\boldsymbol{R}_2}(a\wedge b)$, using the factorization theorem for $\mathbf{R}_3$, Eq.(4.69),the Eq.(4.110) and the multiform identities $B\underset{g}{\times}b = B\underset{g}{\llcorner}b$ and $(B\underset{g}{\llcorner}b)\underset{g}{\cdot}a = B\underset{g}{\cdot}(a\wedge b)$, with $B\in\sec\bigwedge^2 T^*U$ and $a,b\in\sec\bigwedge^1 T^*U$, we have:

$$\begin{aligned}\overset{\varkappa\eta}{\boldsymbol{R}_2}(a\wedge b)\cdot(c\wedge w) &= \overset{\varkappa\eta}{\mathbf{R}_3}(a,b,c,w) \\ &= -\overset{\varkappa\eta}{\rho}(a,b,c)\underset{\eta}{\cdot} w \\ &= -((a\cdot\partial\overset{\varkappa\eta}{\Omega}(b) - b\cdot\partial\overset{\varkappa\eta}{\Omega}(a) - \overset{\varkappa\eta}{\Omega}([a,b]) + \overset{\varkappa\eta}{\Omega}(a)\underset{\eta}{\times}\overset{\varkappa\eta}{\Omega}(b))\underset{\eta}{\times} c)\underset{\eta}{\cdot} w \\ &= (a\cdot\partial\overset{\varkappa\eta}{\Omega}(b) - b\cdot\partial\overset{\varkappa}{\Omega}(a) - \overset{\varkappa\eta}{\Omega}([a,b]) + \overset{\varkappa}{\Omega}(a)\underset{\eta}{\times}\overset{\varkappa\eta}{\Omega}(b))\cdot\underline{\eta}(c\wedge w).\end{aligned}$$

Relabeling $c\to\eta(c)$ and $w\to\eta(w)$, and recalling that $\eta^2 = i_d$, we have

$$\overset{\varkappa\eta}{\boldsymbol{R}_2}(a\wedge b)\cdot\underline{\eta}(c\wedge w) = (a\cdot\partial\overset{\varkappa\eta}{\Omega}(b) - b\cdot\partial\overset{\varkappa\eta}{\Omega}(a) - \overset{\varkappa\eta}{\Omega}([a,b]) + \overset{\varkappa\eta}{\Omega}(a)\underset{\eta}{\times}\overset{\varkappa\eta}{\Omega}(b))\cdot(c\wedge w),$$

which implies that

$$\underline{\eta}\overset{\varkappa\eta}{\boldsymbol{R}_2}(a\wedge b) = a\cdot\partial\overset{\varkappa\eta}{\Omega}(b) - b\cdot\partial\overset{\varkappa\eta}{\Omega}(a) - \overset{\varkappa\eta}{\Omega}([a,b]) + \overset{\varkappa\eta}{\Omega}(a)\underset{\eta}{\times}\overset{\varkappa}{\Omega}(b). \tag{4.111}$$

### *4.8.2 Gauge Extensor Fields of a Lorentz-Cartan MCGSS* $(U_o, \boldsymbol{g}, \gamma)$

We will introduce now three smooth *gauge* extensor fields on $U_o$ obtained from the Minkowski-Cartan *MCGSS* $(U_o,\eta,\varkappa)$, and which encode all the information encoded in the Lorentz-Cartan *MCGSS* $(U_o,\boldsymbol{g},\gamma)$, where $\boldsymbol{g} = \mathbf{h}^\dagger\eta\mathbf{h}$. First, we define

**(i)** The $(2,2)$-extensor field, $B\mapsto\overset{\gamma_h}{\mathcal{R}_2}(B)$, given by

$$\overset{\gamma_h}{\mathcal{R}_2}(B) = \underline{\eta}\overset{\varkappa\eta}{\boldsymbol{R}_2}(B), \tag{4.112}$$

which is called the *Riemann gauge field for* $(U_o,\boldsymbol{g},\gamma)$, and the reason for that name will be clear in a while.

**(ii)** The $(1,1)$-extensor field, $b\mapsto\overset{\gamma_h}{\mathcal{R}_1}(b)$, given by

$$\overset{\gamma_h}{\mathcal{R}_1}(b) = \mathbf{h}^{\clubsuit}(\partial_a)\lrcorner\overset{\gamma_h}{\mathcal{R}_2}(a\wedge b), \tag{4.113}$$

called the Ricci *gauge field for* $(U_o, \boldsymbol{g}, \gamma)$.

**(iii)** The scalar field $\overset{\gamma h}{\mathcal{R}}$, given by by

$$\overset{\gamma h}{\mathcal{R}} = \boldsymbol{h}^{\clubsuit}(\partial_b) \cdot \overset{\gamma h}{\mathcal{R}}_1(b) = \underline{\boldsymbol{h}}^*(\partial_a \wedge \partial_b) \cdot \overset{\gamma h}{\mathcal{R}}_2(a \wedge b), \tag{4.114}$$

called the *gauge Ricci scalar field for* $(U_o, \boldsymbol{g}, \gamma)$.

**(iv)** The $(1,1)$-extensor field, $a \mapsto \overset{\gamma h}{\mathcal{G}}(a)$, given by

$$\overset{\gamma h}{\mathcal{G}}(a) = \overset{\gamma h}{\mathcal{R}}_1(a) - \frac{1}{2}\boldsymbol{h}(a)\overset{\gamma h}{\mathcal{R}}, \tag{4.115}$$

called *the Einstein gauge field for* $(U_o, \boldsymbol{g}, \gamma)$.

**Proposition 4.5** The Riemann gauge field has the fundamental property

$$\overset{\gamma h}{\mathcal{R}}_2(a \wedge b) = a \cdot \partial \overset{\varkappa\eta}{\Omega}(b) - b \cdot \partial \overset{\varkappa\eta}{\Omega}(a) - \overset{\varkappa\eta}{\Omega}([a,b]) + \overset{\varkappa\eta}{\Omega}(a) \underset{\eta}{\times} \overset{\varkappa\eta}{\Omega}(b) \tag{4.116}$$

**Proof** Eq.(4.116) follows immediately from Eq.(4.112) and Eq.(4.111). ∎

**Proposition 4.6** The Ricci scalar field $\overset{\gamma g}{R}$ (associated with $(U_o, \boldsymbol{g}, \gamma)$) is equal to $\overset{\gamma h}{\mathcal{R}}$, i.e.,

$$\overset{\gamma g}{R} = \overset{\gamma h}{\mathcal{R}} \tag{4.117}$$

**Proof** By a simple algebraic manipulation of Eq.(4.70) and using the Theorem 4.1, Eq.(4.109) and Eq.(4.110), we get that

$$\begin{aligned}\overset{\gamma g}{R} &= \underline{g}^{-1}(\partial_a \wedge \partial_b) \cdot \overset{\gamma g}{\boldsymbol{R}}_2(a \wedge b) = \underline{\boldsymbol{h}}^{-1}\underline{\eta}\underline{\boldsymbol{h}}^*(\partial_a \wedge \partial_b) \cdot \underline{\boldsymbol{h}}^\dagger \overset{\varkappa\eta}{\boldsymbol{R}}_2(a \wedge b) \\ &= \underline{\boldsymbol{h}}^*(\partial_a \wedge \partial_b) \cdot \underline{\eta}\overset{\varkappa\eta}{\boldsymbol{R}}_2(a \wedge b) = \underline{\boldsymbol{h}}^*(\partial_a \wedge \partial_b) \cdot \overset{\gamma h}{\mathcal{R}}_2(a \wedge b),\end{aligned}$$

i.e., by Eq.(4.114), $\overset{\gamma g}{R} = \overset{\gamma h}{\mathcal{R}}$. ∎

**Proposition 4.7** The Ricci $(1,1)$-extensor field, $b \mapsto \overset{\gamma g}{\boldsymbol{R}}_1(b)$, and Einstein $(1,1)$-extensor field, $a \mapsto \overset{\gamma g}{\boldsymbol{G}}(a)$, associated to the Lorentz-Cartan *MCGSS* $(U_o, \boldsymbol{g}, \gamma)$ are related to the gauge fields $b \mapsto \overset{\gamma h}{\mathcal{R}}_1(b)$ and $a \mapsto \overset{\gamma h}{\mathcal{G}}(a)$ by

$$\overset{\gamma g}{\boldsymbol{R}}_1(b) = \boldsymbol{h}^\dagger \eta \overset{\gamma h}{\mathcal{R}}_1(b), \tag{4.118}$$

$$\overset{\gamma g}{\boldsymbol{G}}(a) = \boldsymbol{h}^\dagger \eta \overset{\gamma h}{\mathcal{G}}(a). \tag{4.119}$$

**Proof** By simple algebraic manipulation of Eq.(4.70), using Eq.(4.109), Eq. (4.112), Eq.(2.47) and Theorem 4.1, we get that

$$\begin{aligned}\overset{\gamma g}{\boldsymbol{R}}_1(b) &= \boldsymbol{g}^{-1}(\partial_a)\lrcorner\overset{\gamma g}{\boldsymbol{R}}_2(a \wedge b) = \boldsymbol{g}^{-1}(\partial_a)\lrcorner\underline{\boldsymbol{h}}^\dagger \overset{\varkappa\eta}{\boldsymbol{R}}_2(a \wedge b) \\ &= \boldsymbol{g}^{-1}(\partial_a)\lrcorner\underline{\boldsymbol{h}}^\dagger \underline{\eta}\overset{\gamma h}{\mathcal{R}}_2(a \wedge b) = \boldsymbol{h}^\dagger \eta(\eta \boldsymbol{h}\boldsymbol{h}^{-1}\eta \boldsymbol{h}^{\clubsuit}(\partial_a)\lrcorner\overset{\gamma h}{\mathcal{R}}_2(a \wedge b)) \\ &= \boldsymbol{h}^\dagger \eta(\boldsymbol{h}^{\clubsuit}(\partial_a)\lrcorner\overset{\gamma h}{\mathcal{R}}_2(a \wedge b)),\end{aligned}$$

and recalling Eq.(4.113), we have that $\overset{\gamma_g}{\boldsymbol{R}}_1(b) = \boldsymbol{h}^\dagger \eta \overset{\gamma_h}{\mathcal{R}}_1(b)$.

Also, putting Eqs.(4.118) and (4.117) in Eq.(4.73), and using Theorem 4.1, we get

$$\begin{aligned}\overset{\gamma_g}{\boldsymbol{G}}(a) &= \overset{\gamma_g}{\boldsymbol{R}}_1(a) - \frac{1}{2}\boldsymbol{g}(a)\overset{\gamma_g}{R} = \boldsymbol{h}^\dagger \eta \overset{\gamma_h}{\mathcal{R}}_1(a) - \frac{1}{2}\boldsymbol{g}(a)\overset{\gamma_h}{\mathcal{R}} \\ &= \boldsymbol{h}^\dagger \eta \overset{\gamma_h}{\mathcal{R}}_1(a) - \frac{1}{2}\boldsymbol{h}^\dagger \eta \boldsymbol{h}(a)\overset{\gamma_h}{\mathcal{R}} = \boldsymbol{h}^\dagger \eta (\overset{\gamma_h}{\mathcal{R}}(a) - \frac{1}{2}\boldsymbol{h}(a)\overset{\gamma_h}{\mathcal{R}}),\end{aligned}$$

and recalling Eq.(4.115), it follows that $\overset{\gamma_g}{\boldsymbol{G}}(a) = \boldsymbol{h}^\dagger \eta \overset{\gamma_h}{\mathcal{G}}(a)$. ■

The last results may be interpreted by saying that the plastic distortion field $\boldsymbol{h}$ exisitng on the Minkowski *MCGSS* $(U_o, \eta, \mathring{\varkappa})$ deforms its natural parallel transport rule, thus generating $(U_o, \eta, \varkappa)$, a Minkowski-Cartan *MCGSS*[11]. Equivalently, we may say that the field $\boldsymbol{h}$ generates an effective Lorentzian metric (extensor) $\boldsymbol{g}$ and that the Lorentz *MCGSS* $(U_o, \boldsymbol{g}, \gamma)$ is gauge equivalent to the Minkowski-Cartan *MCGSS* $(U_o, \eta, \varkappa)$. This is the case because the Riemann gauge field $\overset{\gamma_h}{\mathcal{R}}(B)$ *encodes all the information contained in the Ricci* $(1,1)$-extensor field $\overset{\gamma_g}{\boldsymbol{R}}_1(a)$, in the Ricci scalar field $\overset{\gamma_g}{R}$ and in the Einstein $(1,1)$-extensor field $\overset{\gamma_g}{\boldsymbol{G}}(a)$ of the Lorentz-Cartan *MCGSS*. Moreover, we have the following proposition.

**Proposition 4.8** The Ricci scalar field $\overset{\gamma_h}{\mathcal{R}}$ $(= \overset{\gamma_g}{R})$ is a *scalar* $\boldsymbol{h}^{\clubsuit}$*-divergent* of the Riemann gauge field $\overset{\gamma_h}{\mathcal{R}}_2(B)$, i.e.,

$$\begin{aligned}\overset{\gamma_h}{\mathcal{R}} &= \underline{\boldsymbol{h}}^*(\partial_a \wedge \partial_b) \cdot \overset{\gamma_h}{\mathcal{R}}_2(a \wedge b) \\ &= \underline{\boldsymbol{h}}^*(\partial_a \wedge \partial_b) \cdot (a \cdot \partial \overset{\varkappa\eta}{\Omega}(b) - b \cdot \partial \overset{\varkappa\eta}{\Omega}(a) - \overset{\varkappa\eta}{\Omega}([a,b]) + \overset{\varkappa\eta}{\Omega}(a) \underset{\eta}{\times} \overset{\varkappa\eta}{\Omega}(b)) \\ \overset{\gamma_h}{\mathcal{R}} &= \underline{\boldsymbol{h}}^*(\partial_a \wedge \partial_b) \cdot ((a \cdot \partial \overset{\varkappa\eta}{\Omega})(b) - (b \cdot \partial \overset{\varkappa\eta}{\Omega})(a) + \overset{\varkappa\eta}{\Omega}(a) \underset{\eta}{\times} \overset{\varkappa\eta}{\Omega}(b)). \qquad (4.120)\end{aligned}$$

The proof of Proposition 4.8 follows trivially from the previous results.

**Remark 4.5** Proposition 4.8 shows that $\overset{\gamma_h}{\mathcal{R}}$ may be interpreted as a *scalar functional* of the plastic gauge distortion field $\boldsymbol{h}$ an the $\varkappa\eta$-gauge rotation field $\overset{\varkappa\eta}{\Omega}$ (and their directional derivatives $\cdot\partial\overset{\varkappa\eta}{\Omega}$). However, to get a simple theory for the gravitational field we need an additional result which is valid for a Lorentz *MCGSS* $(U_o, \boldsymbol{g}, \lambda)$

### *4.8.3 Lorentz MCGSS as* **h***-Deformation of a Particular Minkowski-Cartan MCGSS*

Let $(M, \mathbf{g}, D, \tau_{\mathbf{g}}, \uparrow)$ be a Lorentzian spacetime structure as defined in the beginning of Section 1. We already called the triple $(M, \mathbf{g}, D)$ Lorentz *MCGSS*. As in the previous sections, let $\mathcal{U} \subset M$, $U$ be the canonical vector space, and $U_o \subset U$ the representative of the points of $\mathcal{U}$. Moreover, let $\boldsymbol{g}$ be

[11] The way in which $\mathring{\mathcal{D}}_a$ and $\mathcal{D}_a$ are related is a particular case of the way that two different connections defined on a manifold $M$ are related, see details, e.g., in [4].

the metric extensor on $U_o$ (which represents $\mathbf{g}$) and $\lambda$ the *connection 2-extensor* on $U_o$ representing the de Levi-Civita connection $D$. Under those conditions, we also say that $(U_o, \boldsymbol{g}, \lambda)$ is a Lorentz *MCGSS*.

Observe that the Lorentz $(U_o, \boldsymbol{g}, \lambda)$ *MCGSS* is a particular Lorentz-Cartan *MCGSS* $(U_o, \boldsymbol{g}, \gamma)$ where the connection 2-extensor field is $\gamma = \lambda$ (the Levi-Civita connection 2-extensor field).

Let $D_a$ be the covariant derivative operator defined by $(U_o, \boldsymbol{g}, \lambda)$ and let $a, b \in \sec \bigwedge^1 T^*U$. Under those conditions, we shall use the following *notation* (recall Eq.(4.97))

$$D_a b = a \cdot \partial b + \frac{1}{2}\boldsymbol{g}^{-1}(a \cdot \partial \boldsymbol{g})(b) + \omega(a) \times \boldsymbol{g}(b), \tag{4.121}$$

where the $\boldsymbol{g}$-gauge rotation field $\omega$ of the $(U_o, \boldsymbol{g}, \lambda)$ *GSS* is given by the formula [2]

$$\omega(a) = -\frac{1}{2}\underline{\boldsymbol{g}}^{-1}(\partial_b \wedge \partial_c) A(a, b, c), \tag{4.122}$$

where

$$A(a, b, c) \equiv \frac{1}{2} a \cdot (b \cdot \partial \boldsymbol{g}(c) - c \cdot \partial \boldsymbol{g}(b) - \boldsymbol{g}([b, c])) = \frac{1}{2} a \cdot ((b \cdot \partial \boldsymbol{g})(c) - (c \cdot \partial \boldsymbol{g})(b)), \tag{4.123}$$

is a scalar 3-extensor field (obviously antisymmetric with respect to the second and third variables).

If $\boldsymbol{g} = \boldsymbol{h}^\dagger \eta \boldsymbol{h}$, then, according to the Theorem 4.3 we may interpret the Lorentz *MCGSS* $(U_o, \boldsymbol{g}, \lambda)$ as a deformation of a *particular* Minkowski-Cartan *MCGSS*, which will be denoted by $(U_o, \eta, \mu)$. Indeed, in this case we have the following result:

**Proposition 4.9** [1] If the covariant derivative operator defined by $(U_o, \eta, \mu)$ is denoted $\mathcal{D}_a$ then

$$D_a b = \boldsymbol{h}^{-1}(\mathcal{D}_a \boldsymbol{h}(b)), \tag{4.124}$$

where

$$\mathcal{D}_a b := a \cdot \partial b + \Omega(a) \underset{\eta}{\times} b \tag{4.125}$$

and the $\eta$-gauge rotation operator $\Omega$ of the Minkowski-Cartan $(U_o, \eta, \mu)$ is given by

$$\Omega(a) = -\frac{1}{2}\underline{\eta}^{-1}(\partial_b \wedge \partial_c)[a, \boldsymbol{h}^{-1}(b), \boldsymbol{h}^{-1}(c)], \tag{4.126}$$

where $[a, \boldsymbol{h}^{-1}(b), \boldsymbol{h}^{-1}(c)]$, which has been defined by Eq.(4.42) is the first Christoffel operator associated to the Lorentz metric extensor $\boldsymbol{g} = \boldsymbol{h}^\dagger \eta \boldsymbol{h}$.

**Remark 4.6** It is absolutely clear from the above formulas that the biform field $\Omega(a)$ may be interpreted as a *biform functional* of the plastic distortion gauge field $\boldsymbol{h}$ of its first directional derivatives $\cdot\partial \boldsymbol{h}$ and of its second order directional derivatives $\cdot\partial \cdot \partial \boldsymbol{h}$. It is that result that, summed to the one recalled in Remark 4.5, permits us to formulate a gravitational theory on Minkowski spacetime, where this field is the plastic distortion field $\boldsymbol{h}$. This is, indeed, the case because if $R$ is the curvature scalar in Einstein's *GRT* (the same as $R$ in the Lorentz *GSS* $(U_o, \boldsymbol{g}, \lambda)$), then according to Eq.(4.117) (where now $\gamma = \lambda$), $R \equiv \overset{\lambda_g}{R} = \overset{\lambda_h}{\mathcal{R}}$ is a *scalar functional* of $\boldsymbol{h}^\clubsuit$, $\cdot\partial \boldsymbol{h}^\clubsuit$ and $\cdot\partial \cdot \partial \boldsymbol{h}^\clubsuit$.

# References

[1] Fernández, V. V., Moya, A. M, da Rocha, R., and Rodrigues, W. A. Jr., Clifford and Extensor Calculus and the Riemann and Ricci Extensor Fields of Deformed Structures (M,$\nabla$', $\eta$) and (M,$\nabla$,g), *Int. J. Geom. Meth. Math. Phys.* **4**, 1159-1172 (2007). [http://arxiv.org/abs/math.DG/0502003]
[2] Fernández, V. V., Moya, A. M., and Rodrigues, W. A. Jr., Geometric and and Extensor Algebras in the Study of the Differential Geometry of Arbitrary Manifolds, *Int. J. Geom. Meth. Math. Phys.* **4**, 1117-1158 (2007). [http://arxiv.org/abs/math.DG/0703094]
[3] Choquet-Bruhat, Y., DeWitt-Morette, C., and Dillard-Bleick, M., *Analysis, Manifolds and Physics* (revised edition), North Holland Publ. Co., Amsterdam, 1982.
[4] Rodrigues, W. A. Jr., and Capelas de Oliveira, E., T*he Many Faces of Maxwell, Dirac and Einstein Equations. A Clifford Bundle Approach.* Lecture Notes in Physics **722**, Springer, Heidelberg, 2007.
[5] Zorawski, M., *Theorie Mathematiques des Dislocations*, Dunod, Paris, 1967.
[6] Frankel, T., *The Geometry of Physics*, (revised edition), Cambridge Univ. Press, Cambridige, 1997.
[7] Notte-Cuello, E. A., and Rodrigues, W. A. Jr., A Maxwell Like Formulation of Gravitational Theory in Minkowski Spacetime, *Int. J. Mod. Phys. D* **16**, 1027-1041 (2007). [http://arxiv.org/abs/math-ph/0608017]

# 5

# Gravitation as Plastic Distortion of the Lorentz Vacuum

ABSTRACT In this chapter we will introduce some notation in Section 5.1 and, then, in Section 5.2 we will present the Lagrangian for the free gravitational field $\boldsymbol{h}^{\clubsuit}$ : $\sec \bigwedge^1 T^*U \to \sec \bigwedge^1 T^*U$. In Section 5.3, dedicated to the derivation of the equation of motion for $\boldsymbol{h}^{\clubsuit}$ that follows from a variational principle, we will see the power of the mathematical methods developed in Chapters 2, 3 and 4. In Section 5 we will present the Lagrangian for the gravitational field plus a matter field which includes also a cosmological constant term and, then, will derive the equations of motion for the gravitational field.

## 5.1 Notation for This Chapter

The Remark 4.6 clearly reveals the way that the dynamics of the plastic gauge deformations $\boldsymbol{h}^{\clubsuit}$ of the Lorentz vacuum must be described in a world that unless experimental facts (and none exists now, for the best of our knowledge) demonstrate the contrary, must be described by an event manifold $M \simeq \mathbb{R}^4$. We elaborate this point as follows.

Let $(M \simeq \mathbb{R}^4, \boldsymbol{\eta}, D, \tau_\eta, \uparrow)$ be the structure representing Minkowski spacetime as defined in Chapter 1. In our theory, we suppose that all physical fields and/or particles live and interact in the arena defined by Minkowski spacetime. Now, let $(M \simeq \mathbb{R}^4, \mathbf{g}, D, \tau_{\mathbf{g}}, \uparrow)$ be a Lorentzian manifold that as we know represents a particular gravitational field in Einstein's *GRT*. Moreover, let $\eta'$,$\mathbf{g} \in \sec T_0^2 M$ be the metric tensors on the cotangent bundle such that, in the global chart with global coordinates [1] $\{\mathrm{x}^\alpha\}$ associated to a chart $(M, \phi)_o$, where

$$\boldsymbol{\eta} = \eta_{\alpha\beta} d\mathrm{x}^\alpha \otimes d\mathrm{x}^\beta, \quad \mathbf{g} = g_{\alpha\beta} d\mathrm{x}^\alpha \otimes d\mathrm{x}^\beta \tag{5.1}$$

[1]Called coordinates in the Einstein-Lorentz-Poincaré gauge.

V.V. Fernández and W.A. Rodrigues Jr., *Gravitation as a Plastic Distortion of the Lorentz Vacuum*, Fundamental Theories of Physics 168, DOI 10.1007/978-3-642-13589-7_5,

we have

$$\eta' = \eta^{\alpha\beta}\frac{\partial}{\partial \mathrm{x}^{\alpha}} \otimes \frac{\partial}{\partial \mathrm{x}^{\beta}}, \qquad \mathtt{g} = g^{\alpha\beta}\frac{\partial}{\partial \mathrm{x}^{\alpha}} \otimes \frac{\partial}{\partial \mathrm{x}^{\beta}} \tag{5.2}$$

with $\eta_{\alpha\beta}\eta^{\beta\nu} = \delta^{\nu}_{\alpha}$ and $g_{\alpha\beta}g^{\beta\nu} = \delta^{\nu}_{\alpha}$. Let also $\eta$ be the extensor field corresponding to $\eta$, i.e.,

$$\eta'(d\mathrm{x}^{\alpha}, d\mathrm{x}^{\beta}) = \eta(d\mathrm{x}^{\alpha}) \cdot d\mathrm{x}^{\beta} \tag{5.3}$$

where in Eq.(5.3) the symbol $\cdot$ denotes the canonical scalar product on $M \simeq \mathbb{R}^4$, i.e., $d\mathrm{x}^{\alpha}\cdot d\mathrm{x}^{\beta} = \delta^{\alpha\beta}$.

Since $M \simeq \mathbb{R}^4$, using the global coordinates $\{\mathrm{x}^{\alpha}\}$, we have immediately that $M \simeq U_o \simeq U$ (where $U_o$ and $U$ are as previously introduced) and the equivalence classes of the $d\mathrm{x}^{\alpha}$ are $\boldsymbol{\vartheta}^{\alpha} := [d\mathrm{x}^{\alpha}]$, which $\{\boldsymbol{\vartheta}^{\alpha}\}$ defining a basis[2] for $U$.

Its reciprocal basis will be denoted by $\{\vartheta_{\alpha}\}$. In this case, given a smooth invertible extensor field $\mathbf{h} : \sec \bigwedge^1 TM \to \sec \bigwedge^1 TM$ its representative on $U$ is the smooth extensor field $\boldsymbol{h} : \sec \bigwedge^1 T^*U \to \sec \bigwedge^1 T^*U$. Recall, moreover, that the metric tensor $\mathbf{g}$ is represented on $U$ by the extensor field $\boldsymbol{g} : \sec \bigwedge^1 T^*U \to \sec \bigwedge^1 T^*U$,

$$\boldsymbol{g} = \boldsymbol{h}^{\dagger}\eta\boldsymbol{h}, \tag{5.4}$$

whereas the representative of $\mathtt{g}$ is

$$\boldsymbol{g}^{-1} = \boldsymbol{h}^{-1}\eta\boldsymbol{h}^{-1\dagger} \tag{5.5}$$

with (for $\mathbf{x} \in U$ and $x \in M$, $\mathbf{x} = \phi(x)$)

$$\boldsymbol{g}^{-1}\big|_{\mathbf{x}}(\boldsymbol{\vartheta}^{\alpha}) \cdot \boldsymbol{\vartheta}^{\beta} := \mathtt{g}(d\mathrm{x}^{\alpha}, d\mathrm{x}^{\beta})\big|_{x}, \tag{5.6}$$

In what follows, we present a gravitational theory where that field which exists in Minkowski spacetime is represented by smooth invertible extensor field $\boldsymbol{h} : \sec \bigwedge^1 T^*U \to \sec \bigwedge^1 T^*U$ describing (where it is *not* the identity) a plastic deformation of the Lorentz vacuum, putting it in an 'excited' state which may be described by Minkowski-Cartan *MCGSS* $(M, \eta, \varkappa)$, which, as we already know, is gauge equivalent to the Lorentz *MCGSS* $(M, \boldsymbol{g}, \gamma)$. We will give the Lagrangian encoding the dynamics of $\boldsymbol{h}$ and its interaction with the matter fields and will derive its equation of motion.

## 5.2 Lagrangian for the Free $\boldsymbol{h}^{\clubsuit}$ Field

Under the conditions established above we postulate that the dynamics of the gravitational field $\boldsymbol{h} : \sec \bigwedge^1 T^*U \to \sec \bigwedge^1 T^*U$ is encoded in the following Lagrangian, which using notations introduced in the previous section, is written as[3]

$$[\boldsymbol{h}^{\clubsuit}, \cdot\partial\boldsymbol{h}^{\clubsuit}, \cdot\partial\cdot\partial\boldsymbol{h}^{\clubsuit}] \mapsto \mathfrak{L}_{eh}[\boldsymbol{h}^{\clubsuit}, \cdot\partial\boldsymbol{h}^{\clubsuit}, \cdot\partial\cdot\partial\boldsymbol{h}^{\clubsuit}] := \frac{1}{2}\mathcal{R}\det[\boldsymbol{h}], \tag{5.7}$$

where recalling Eq.(4.120) and Eq.(4.126)) we see that $\mathcal{R} \equiv \overset{\lambda_h}{\mathcal{R}}$ may indeed be expressed as a scalar functional field of extensor variables $\boldsymbol{h}^{\clubsuit}$, $\cdot\partial\boldsymbol{h}^{\clubsuit}$, and $\cdot\partial\cdot\partial\boldsymbol{h}^{\clubsuit}$, i.e.,

$$\begin{aligned}\mathcal{R} &= \underline{\boldsymbol{h}}^*(\partial_a \wedge \partial_b) \cdot \mathcal{R}_2(a \wedge b) \\ &= \underline{\boldsymbol{h}}^*(\partial_a \wedge \partial_b) \cdot (a \cdot \partial(\Omega b) - b \cdot \partial\Omega(a) - \Omega([a,b]) + \Omega(a) \underset{\eta}{\times} \Omega(b)),\end{aligned} \tag{5.8}$$

[2] The canonical reciprocal basis of $\{\boldsymbol{\vartheta}^{\alpha}\}$ of $U$ is denoted by $\{\vartheta_{\alpha}\}$. The basis $\{\boldsymbol{\vartheta}_{\alpha}\}$ of $U$ is defined by the condition $\boldsymbol{\vartheta}^{\alpha} \underset{\eta^{-1}}{\cdot} \boldsymbol{\vartheta}_{\beta} = \boldsymbol{\vartheta}^{\alpha} \underset{\eta}{\cdot} \boldsymbol{\vartheta}_{\beta} = \delta^{\alpha}_{\beta}$. It may be called the $\eta$-reciprocal basis of $\{\boldsymbol{\vartheta}^{\alpha}\}$.

[3] The notation $\mathfrak{L}_{eh}$ is used because it is equivalent to the Einstein-Hilbert Lagrangian.

where the biform field $\Omega(a)$ given by Eq.(4.122), is also expressed as a functional for the extensor fields $\boldsymbol{h}^{\clubsuit}$, $\cdot\partial\boldsymbol{h}^{\clubsuit}$ and $\cdot\partial\cdot\partial\boldsymbol{h}^{\clubsuit}$.

To continue, we must express $\det[\boldsymbol{h}]$ as a scalar functional of $\boldsymbol{h}^{\clubsuit}$also, i.e.,

$$\det[\boldsymbol{h}] = \frac{1}{\det[\boldsymbol{h}^{\clubsuit}]}. \tag{5.9}$$

The action functional for the field $\boldsymbol{h}^{\clubsuit}$ on $U$ is

$$\begin{aligned} \mathcal{A} &= \int_U \mathfrak{L}_{eh}[\boldsymbol{h}^{\clubsuit}, \cdot\partial\boldsymbol{h}^{\clubsuit}, \cdot\partial\cdot\partial\boldsymbol{h}^{\clubsuit}]\ \tau \\ &= \frac{1}{2}\int_U \mathcal{R}\det[\boldsymbol{h}]\ \tau, \end{aligned} \tag{5.10}$$

where $\tau$ is an arbitrary Euclidean volume element, e.g., we can take $\tau = dx^0 \wedge dx^1 \wedge dx^2 \wedge dx^3$

Let $\boldsymbol{\delta}^{w}_{h^{\clubsuit}}$ be the variational operator with respect to $\boldsymbol{h}^{\clubsuit}$ in the direction of the smooth $(1,1)$-extensor field $w$, such that $w|_{\partial U} = 0$ and $\cdot\partial w|_{\partial U} = 0$.

## 5.3 Equation of Motion for $\boldsymbol{h}^{\clubsuit}$

As usual, in the Lagrangian formalism we suppose that the dynamics of $\boldsymbol{h}^{\clubsuit}$is given by the principle of stationary action,

$$\boldsymbol{\delta}^{w}_{h^{\clubsuit}} \int_U \mathfrak{L}_{eh}[\boldsymbol{h}^{\clubsuit}, \cdot\partial\boldsymbol{h}^{\clubsuit}, \cdot\partial\cdot\partial\boldsymbol{h}^{\clubsuit}]\tau = \int_U \boldsymbol{\delta}^{w}_{h^{\clubsuit}}(\mathcal{R}\det[\boldsymbol{h}])\ \tau = 0, \tag{5.11}$$

which implies a functional differential equation (Euler-Lagrange equation) for $\boldsymbol{h}^{\clubsuit}$ that we now derive.

As is well known the solution of that problem implies the use of variational formulas, the Gauss-Stokes theorem (for star shape regions), i.e.,

$$\int_U \partial\cdot a\ \tau = \int_{\partial U} \vartheta^{\mu}\cdot a\ \tau_{\mu}, \tag{5.12}$$

where $\tau_{\mu}$ are 3-form fields on the boundary $\partial U$ of $U$, $a$ is a smooth 1-form field and also a fundamental lemma of integration theory, which says that if $\int_U A\cdot X\ \tau = 0$ for all multiform field $A$ then the $X = 0$.

Before starting the calculations, we recall from Section 3.8 that any variational operator $\boldsymbol{\delta}^{w}_{t}$ satisfies the following rules. Let $\Phi, \Psi$ be arbitrary multiform functionals of the extensor field $t$. Moreover, let $F$ be a scalar functional of the extensor field $t$ with $a$ a smooth 1-form field and $\varphi$ an arbitrary function. Then:

$$\boldsymbol{\delta}^{w}_{t}(\Phi[t] * \Psi[t]) = (\boldsymbol{\delta}^{w}_{t}\Phi[t]) * \Psi[t] + \Phi[t] * (\boldsymbol{\delta}^{w}_{t}\Psi[t]), \tag{5.13a}$$

$$\boldsymbol{\delta}^{w}_{t} t(a) = w(a), \tag{5.13b}$$

$$\boldsymbol{\delta}^{w}_{t}(a\cdot\partial\Phi[t]) = a\cdot\partial(\boldsymbol{\delta}^{w}_{t}\Phi[t]), \tag{5.13c}$$

$$\boldsymbol{\delta}^{w}_{t}\varphi(F[t]) = \varphi'(F[t])(\boldsymbol{\delta}^{w}_{t}F[t]), \tag{5.13d}$$

where in Eq.(5.13a) $*$ denotes (as previously agreed) any of the multiform products.

Using Eq.(5.13a) in Eq.(5.7), and taking into account Eq.(5.8), we get

$$
\begin{aligned}
\boldsymbol{\delta}^{w}_{h^\clubsuit}(\underline{\boldsymbol{h}}^\clubsuit(\partial_a \wedge \partial_b) \cdot \mathcal{R}_2(a \wedge b) \det[\boldsymbol{h}]) &= (\boldsymbol{\delta}^{w}_{h^\clubsuit}\underline{\boldsymbol{h}}^\clubsuit(\partial_a \wedge \partial_b)) \cdot \mathcal{R}_2(a \wedge b) \det[\boldsymbol{h}] \\
&+ \underline{\boldsymbol{h}}^\clubsuit(\partial_a \wedge \partial_b) \cdot (\boldsymbol{\delta}^{w}_{\boldsymbol{h}^\clubsuit}\mathcal{R}_2(a \wedge b)) \det[\boldsymbol{h}] \\
&+ \underline{\boldsymbol{h}}^\clubsuit(\partial_a \wedge \partial_b) \cdot \mathcal{R}_2(a \wedge b)(\boldsymbol{\delta}^{w}_{h^\clubsuit} \det[\boldsymbol{h}]).
\end{aligned} \tag{5.14}
$$

Next we will obtain a functional identity for the scalar product in the first term of Eq.(5.14). We have,

$$
\begin{aligned}
&(\boldsymbol{\delta}^{w}_{h^\clubsuit}\underline{\boldsymbol{h}}^\clubsuit(\partial_a \wedge \partial_b)) \cdot \mathcal{R}_2(a \wedge b) \\
&= (\boldsymbol{\delta}^{w}_{\boldsymbol{h}^\clubsuit}\boldsymbol{h}^\clubsuit(\partial_a) \wedge \boldsymbol{h}^\clubsuit(\partial_b) + \boldsymbol{h}^\clubsuit(\partial_a) \wedge \boldsymbol{\delta}^{w}_{h^\clubsuit}\boldsymbol{h}^\clubsuit(\partial_b)) \cdot \mathcal{R}_2(a \wedge b) \\
&= (w(\partial_a) \wedge \boldsymbol{h}^\clubsuit(\partial_b) + \boldsymbol{h}^\clubsuit(\partial_a) \wedge w(\partial_b)) \cdot \mathcal{R}_2(a \wedge b) \\
&= (w(\partial_a) \wedge \boldsymbol{h}^\clubsuit(\partial_b)) \cdot \mathcal{R}_2(a \wedge b) + (\boldsymbol{h}^\clubsuit(\partial_a) \wedge w(\partial_b)) \cdot \mathcal{R}_2(a \wedge b) \\
&= (\boldsymbol{h}^\clubsuit(\partial_b) \wedge w(\partial_a)) \cdot \mathcal{R}_2(b \wedge a) + (\boldsymbol{h}^\clubsuit(\partial_b) \wedge w(\partial_a)) \cdot \mathcal{R}_2(b \wedge a) \\
&= 2(\boldsymbol{h}^\clubsuit(\partial_b) \wedge w(\partial_a)) \cdot \mathcal{R}_2(b \wedge a) \\
&= 2w(\partial_a) \cdot (\boldsymbol{h}^\clubsuit(\partial_b) \lrcorner \mathcal{R}_2(b \wedge a)),
\end{aligned}
$$

i.e.,

$$
(\boldsymbol{\delta}^{w}_{h^\clubsuit}\underline{\boldsymbol{h}}^\clubsuit(\partial_a \wedge \partial_b)) \cdot \mathcal{R}_2(a \wedge b) = 2w(\partial_a) \cdot \mathcal{R}_1(a), \tag{5.15}
$$

where we used Eq.(5.13a) in the form $\boldsymbol{\delta}^{w}_{t}(\boldsymbol{\Phi}[t] \wedge \boldsymbol{\Psi}[t]) = (\boldsymbol{\delta}^{w}_{t}\boldsymbol{\Phi}[t]) \wedge \boldsymbol{\Psi}[t] + \boldsymbol{\Phi}[t]) \wedge (\boldsymbol{\delta}^{w}_{t}\boldsymbol{\Psi}[t])$, Eq.(5.13b) and Eq.(4.113) which defines the Ricci gauge field.

Next we will obtain a second functional identity for the scalar product in the second term[4] of Eq.(5.14),

$$
\begin{aligned}
\underline{\boldsymbol{h}}^\clubsuit(\partial_a \wedge \partial_b) \cdot \boldsymbol{\delta}^{w}_{h^\clubsuit}\mathcal{R}_2(a \wedge b) &= \underline{\boldsymbol{h}}^\clubsuit(\boldsymbol{\vartheta}^\mu \wedge \boldsymbol{\vartheta}^\nu) \cdot \boldsymbol{\delta}^{w}_{h^\clubsuit}\mathcal{R}_2(\vartheta_\mu \wedge \vartheta_\nu) \\
&= \underline{\boldsymbol{h}}^\clubsuit(\boldsymbol{\vartheta}^\mu \wedge \boldsymbol{\vartheta}^\nu) \cdot \boldsymbol{\delta}^{w}_{h^\clubsuit}(\vartheta_\mu \cdot \partial\Omega(\vartheta_\nu) \\
&- \vartheta_\nu \cdot \partial\Omega(\vartheta_\mu) + \Omega(\vartheta_\mu) \underset{\eta}{\times} \Omega(\vartheta_\nu)) \\
&= \underline{\boldsymbol{h}}^\clubsuit(\boldsymbol{\vartheta}^\mu \wedge \boldsymbol{\vartheta}^\nu) \cdot (\vartheta_\mu \cdot \partial\boldsymbol{\delta}^{w}_{h^\clubsuit}\Omega(\vartheta_\nu)\Omega(\vartheta_\mu) \\
&- \vartheta_\nu \cdot \partial\boldsymbol{\delta}^{w}_{h^\clubsuit} \\
&+ \delta^{w}_{\boldsymbol{h}^\clubsuit}\Omega(\vartheta_\mu) \underset{\eta}{\times} \Omega(\vartheta_\nu) + \Omega(\vartheta_\mu) \underset{\eta}{\times} \boldsymbol{\delta}^{w}_{\boldsymbol{h}^\clubsuit}\Omega(\vartheta_\nu)) \\
&= \underline{\boldsymbol{h}}^\clubsuit(\boldsymbol{\vartheta}^\mu \wedge \boldsymbol{\vartheta}^\nu) \cdot (\mathcal{D}_{\vartheta_\mu}\boldsymbol{\delta}^{w}_{\boldsymbol{h}^\clubsuit}\Omega(\vartheta_\nu) \\
&- \mathcal{D}_{\vartheta_\nu}\boldsymbol{\delta}^{w}_{\boldsymbol{h}^\clubsuit}\Omega(\vartheta_\mu)) \\
&= 2\underline{\boldsymbol{h}}^\clubsuit(\boldsymbol{\vartheta}^\mu \wedge \boldsymbol{\vartheta}^\nu) \cdot \mathcal{D}_{\vartheta_\mu}\delta^{w}_{h^\clubsuit}\Omega(\vartheta_\nu) \\
&= 2\boldsymbol{h}^\clubsuit(\vartheta^\nu) \cdot (\boldsymbol{h}^\clubsuit(\vartheta^\mu) \lrcorner \mathcal{D}_{\vartheta_\mu}\delta^{w}_{h^\clubsuit}\Omega(\vartheta_\nu)) \\
&= 2\boldsymbol{h}^\clubsuit(\vartheta^\nu) \cdot \boldsymbol{D} \lrcorner \boldsymbol{\delta}^{w}_{h^\clubsuit}\Omega(\vartheta_\nu).
\end{aligned} \tag{5.16}
$$

[4]Keep in mind that $\{\vartheta_\mu\}$ is the canonical reciprocal basis of $\{\boldsymbol{\vartheta}^\mu\}$, i.e., $\boldsymbol{\vartheta}^\mu \cdot \vartheta_\nu = \delta^\mu_\nu$.

In these equations, we essentially used Eq.(5.13c) once more and Eq.(5.13a), this time in the form, $\boldsymbol{\delta}_t^w(\boldsymbol{\Phi}[t] \underset{\eta}{\times} \boldsymbol{\Psi}[t]) = (\boldsymbol{\delta}_t^w\boldsymbol{\Phi}[t]) \underset{\eta}{\times} \boldsymbol{\Psi}[t] + \boldsymbol{\Phi}[t]) \underset{\eta}{\times} (\boldsymbol{\delta}_t^w\boldsymbol{\Psi}[t])$, and the definition of the gauge divergent $\boldsymbol{D}\lrcorner X = \boldsymbol{h}^{\clubsuit}(\partial_a)\lrcorner\mathcal{D}_a X$, where $\mathcal{D}_a$ is the covariant derivative operator, which applied on a smooth multiform field $X$ gives

$$\mathcal{D}_a X := a \cdot \partial X + \Omega(a) \underset{\eta}{\times} X. \tag{5.17}$$

We may also write the right hand side of Eq.(5.12), which is a scalar product as the scalar divergent of a smooth 1-form field, using the following identity (which is easily obtained if we take Eq.(4.87) into account),

$$\boldsymbol{D}\lrcorner X = \frac{1}{\det[\boldsymbol{h}]}\underline{\boldsymbol{h}}(\partial\lrcorner \det[\boldsymbol{h}]\underline{\boldsymbol{h}}^{-1}(X)). \tag{5.18}$$

We then have

$$\begin{aligned}
\boldsymbol{h}^{\clubsuit}(\boldsymbol{\vartheta}^{\nu}) \cdot \boldsymbol{D}\lrcorner\boldsymbol{\delta}^w_{h^{\clubsuit}}\Omega(\vartheta_\nu) &= \boldsymbol{h}^{\clubsuit}(\boldsymbol{\vartheta}^{\nu}) \cdot \frac{1}{\det[\boldsymbol{h}]}\boldsymbol{h}(\partial\lrcorner \det[\boldsymbol{h}]\underline{\boldsymbol{h}}^{-1}(\boldsymbol{\delta}^w_{h^{\clubsuit}}\Omega(\vartheta_\nu)))\\
&= \frac{1}{\det[\boldsymbol{h}]}\boldsymbol{\vartheta}^{\nu} \cdot \partial\lrcorner(\det[\boldsymbol{h}]\underline{\boldsymbol{h}}^{-1}(\boldsymbol{\delta}^w_{h^{\clubsuit}}\Omega(\vartheta_\nu))\\
&= \frac{1}{\det[\boldsymbol{h}]}\boldsymbol{\vartheta}^{\nu} \cdot (\boldsymbol{\vartheta}^{\mu}\lrcorner\vartheta_\mu \cdot \partial(\det[\boldsymbol{h}]\underline{\boldsymbol{h}}^{-1}(\boldsymbol{\delta}^w_{h^{\clubsuit}}\Omega(\vartheta_\nu)))\\
&= \frac{1}{\det[\boldsymbol{h}]}(\boldsymbol{\vartheta}^{\mu} \wedge \boldsymbol{\vartheta}^{\nu}) \cdot \vartheta_\mu \cdot \partial(\det[\boldsymbol{h}]\underline{\boldsymbol{h}}^{-1}(\boldsymbol{\delta}^w_{h^{\clubsuit}}\Omega(\vartheta_\nu))\\
&= -\frac{1}{\det[\boldsymbol{h}]}(\boldsymbol{\vartheta}^{\mu} \wedge \boldsymbol{\vartheta}^{\nu}) \cdot \vartheta_\mu \cdot \partial(\det[\boldsymbol{h}]\underline{\boldsymbol{h}}^{-1}(\boldsymbol{\delta}^w_{h^{\clubsuit}}\Omega(\vartheta_\nu))\\
&= -\frac{1}{\det[\boldsymbol{h}]}\boldsymbol{\vartheta}^{\mu} \cdot (\boldsymbol{\vartheta}^{\nu}\lrcorner\vartheta_\mu \cdot \partial(\det[\boldsymbol{h}]\underline{\boldsymbol{h}}^{-1}(\boldsymbol{\delta}^w_{h^{\clubsuit}}\Omega(\vartheta_\nu)))\\
&= -\frac{1}{\det[\boldsymbol{h}]}\boldsymbol{\vartheta}^{\mu} \cdot \vartheta_\mu \cdot \partial(\boldsymbol{\vartheta}^{\nu}\lrcorner \det[\boldsymbol{h}]\underline{\boldsymbol{h}}^{-1}(\boldsymbol{\delta}^w_{h^{\clubsuit}}\Omega(\vartheta_\nu)))\\
&= -\frac{1}{\det[\boldsymbol{h}]}\partial \cdot (\det[\boldsymbol{h}]\boldsymbol{\vartheta}^{\nu}\lrcorner\underline{\boldsymbol{h}}^{-1}(\boldsymbol{\delta}^w_{h^{\clubsuit}}\Omega(\vartheta_\nu))).
\end{aligned}$$

i.e.,

$$\boldsymbol{h}^{\clubsuit}(\boldsymbol{\vartheta}^{\nu}) \cdot \boldsymbol{D}\lrcorner\boldsymbol{\delta}^w_{h^{\clubsuit}}\Omega(\vartheta_\nu) = -\frac{1}{\det[\boldsymbol{h}]}\partial \cdot (\det[\boldsymbol{h}]\partial_a\lrcorner\underline{\boldsymbol{h}}^{-1}(\boldsymbol{\delta}^w_{h^{\clubsuit}}\Omega(a))). \tag{5.19}$$

Now, putting Eq.(5.19) in Eq.(5.16), we get an analogy of the well known *Palatini identity,* i.e.,

$$\underline{\boldsymbol{h}}^{\clubsuit}(\partial_a \wedge \partial_b) \cdot \boldsymbol{\delta}^w_{h^{\clubsuit}}\mathcal{R}_2(a \wedge b) = -2\frac{1}{\det[\boldsymbol{h}]}\partial \cdot (\det[\boldsymbol{h}]\partial_a\lrcorner\underline{\boldsymbol{h}}^{-1}(\boldsymbol{\delta}^w_{h^{\clubsuit}}\Omega(a))). \tag{5.20}$$

Next, we calculate the variation of $\det[\boldsymbol{h}]$ with respect to $\boldsymbol{h}^{\clubsuit}$ in the direction of $w$. In order to do so, we use the chain rule given by Eq.(5.13d) and the variational formula (recall Eq.(3.68))

$$\boldsymbol{\delta}_t^w \det[t] = w(\partial_a) \cdot t^{\clubsuit}(a) \det[t], \tag{5.21}$$

which permits us to write

$$\begin{aligned}
\boldsymbol{\delta}^{w}_{h^\clubsuit} \det[\boldsymbol{h}] &= \boldsymbol{\delta}^{w}_{h^\clubsuit} \frac{1}{\det[\boldsymbol{h}^\clubsuit]} \\
&= -\frac{1}{(\det[\boldsymbol{h}^\clubsuit])^2} \boldsymbol{\delta}^{w}_{h^\clubsuit} \det[\boldsymbol{h}^\clubsuit] \\
&= -\frac{1}{(\det[\boldsymbol{h}^\clubsuit])^2} w(\partial_a) \cdot \boldsymbol{h}(a) \det[\boldsymbol{h}^\clubsuit] \\
&= -\frac{1}{\det[\boldsymbol{h}^\clubsuit]} w(\partial_a) \cdot \boldsymbol{h}(a) \\
\boldsymbol{\delta}^{w}_{h^\clubsuit} \det[\boldsymbol{h}] &= -w(\partial_a) \cdot \boldsymbol{h}(a) \det[\boldsymbol{h}]. \qquad (5.22)
\end{aligned}$$

Finally, using Eqs.(5.15), (5.20) and (5.22) in Eq.(5.14) we get the variation with respect to $\boldsymbol{h}^\clubsuit$ in the direction of $w$ of the Lagrangian given by Eq.(5.7),

$$\begin{aligned}
\boldsymbol{\delta}^{w}_{h^\clubsuit}(\mathcal{R} \det[\boldsymbol{h}]) &= 2w(\partial_a) \cdot \mathcal{R}_1(a) \det[\boldsymbol{h}] \\
&\quad - 2\frac{1}{\det[\boldsymbol{h}]} \partial \cdot (\det[\boldsymbol{h}] \partial_a \lrcorner \underline{\boldsymbol{h}}^{-1}(\boldsymbol{\delta}^{w}_{h^\clubsuit} \Omega(a))) \det[\boldsymbol{h}] \\
&\quad - \underline{\boldsymbol{h}}^\clubsuit(\partial_a \wedge \partial_b) \cdot \mathcal{R}_2(a \wedge b)(w(\partial_a) \cdot \boldsymbol{h}(a) \det[\boldsymbol{h}]) \\
&= 2w(\partial_a) \cdot (\mathcal{R}_1(a) - \frac{1}{2} \boldsymbol{h}(a) \mathcal{R}) \det[\boldsymbol{h}]) \\
&\quad - 2w(\partial_a) \cdot (2\partial \cdot (\det[\boldsymbol{h}] \partial_a \lrcorner \underline{\boldsymbol{h}}^{-1}(\boldsymbol{\delta}^{w}_{\boldsymbol{h}^\clubsuit} \Omega(a))) \\
\boldsymbol{\delta}^{w}_{h^\clubsuit}(\mathcal{R} \det[\boldsymbol{h}]) &= 2w(\partial_a) \cdot \mathcal{G}(a) \det[\boldsymbol{h}] - 2\partial \cdot (\det[\boldsymbol{h}] \partial_a \lrcorner \underline{\boldsymbol{h}}^{-1}(\boldsymbol{\delta}^{w}_{h^\clubsuit} \Omega(a))). \qquad (5.23)
\end{aligned}$$

In the last step we recalled the definition (Eq.(4.115)) of the Einstein gauge field $\mathcal{G}$ for the particular case of the Lorentz *MCGSS*.

Taking into account Eq.(5.23), the contour problem for the dynamic variable $\boldsymbol{h}^\clubsuit$ becomes

$$\int_U w(\partial_a) \cdot \mathcal{G}(a) \det[\boldsymbol{h}] \; \tau - \int_U \partial \cdot (\det[\boldsymbol{h}] \partial_a \lrcorner \underline{\boldsymbol{h}}^{-1}(\boldsymbol{\delta}^{w}_{h^\clubsuit} \Omega(a))) \; \tau = 0, \qquad (5.24)$$

for all $w$ satisfying the boundary conditions $w|_{\partial U} = 0$ and $a \cdot \partial w|_{\partial U} = 0$.

Then, using the Gauss-Stokes theorem with the above boundary conditions the second term of Eq.(5.24) may be integrated and gives

$$\begin{aligned}
&\int_U \partial \cdot (\det[\boldsymbol{h}] \partial_a \lrcorner \underline{\boldsymbol{h}}^{-1}(\boldsymbol{\delta}^{w}_{h^\clubsuit} \Omega(a))) \; \tau \\
&= \int_{\partial U} \det[\boldsymbol{h}] \vartheta^\mu \cdot (\partial_a \lrcorner \underline{\boldsymbol{h}}^{-1}(\boldsymbol{\delta}^{w}_{h^\clubsuit} \Omega(a))) \; \tau_\mu = 0, \qquad (5.25)
\end{aligned}$$

since under the boundary conditions $w|_{\partial U} = 0$ and $a \cdot \partial w|_{\partial U} = 0$, it is $\boldsymbol{\delta}^{w}_{\boldsymbol{h}^\clubsuit} \Omega(a) = 0$.

Thus, using Eq.(5.25) in Eq.(5.24) it follows that

$$\int_U w(\partial_a) \cdot \mathcal{G}(a) \det[\boldsymbol{h}] \; \tau = 0, \qquad (5.26)$$

for all $w$, and since $w$ is arbitrary, a fundamental lemma of integration theory yields

$$\mathcal{G}(a)\det[\boldsymbol{h}] = 0, \tag{5.27}$$

i.e.,

$$\mathcal{G}(a) = \mathcal{R}_1(a) - \frac{1}{2}\boldsymbol{h}(a)\mathcal{R} = 0, \tag{5.28}$$

which is the field equation for the distortion gauge field $\boldsymbol{h}$.

If we recall the relation between the Einstein $(1,1)$-extensor field, $a \mapsto \boldsymbol{G}(a)$ and the Einstein gauge field, $a \mapsto \mathcal{G}(a)$, (i.e., $\boldsymbol{G}(a) = \boldsymbol{h}^{\dagger}\eta\mathcal{G}(a)$), we get

$$\boldsymbol{G}(a) = 0, \tag{5.29}$$

which we recognize as equivalent to the free Einstein equation for $\boldsymbol{g}$.

## 5.4 Lagrangian for the Gravitational Field Plus Matter Field Including a Cosmological Constant Term

Since cosmological data seems to suggest an expansion of the universe (when it is described in terms of the Lorentzian spacetime model), we postulate here that the total Lagrangian describing the interaction of the gravitational field with matter is[5]

$$\mathfrak{L} = \mathfrak{L}_{eh} + \lambda\det[\boldsymbol{h}] + \mathfrak{L}_m \det[\boldsymbol{h}], \tag{5.30}$$

where $\lambda$ is the cosmological constant and where $\mathfrak{L}_m$ is the matter Lagrangian. The equations of motion for $\boldsymbol{h}$ are obtained from the variational principle,

$$\int_U \boldsymbol{\delta}^w_{h^\clubsuit}(\frac{1}{2}\mathcal{R} + \lambda + \mathfrak{L}_m)\det[\boldsymbol{h}]\ \tau = 0. \tag{5.31}$$

Then, we get, defining conveniently the energy-momentum extensor of matter $T(a)$ by

$$(w(\partial_a)\cdot\eta\boldsymbol{h}^{\clubsuit}T(a))\det[\boldsymbol{h}] := \boldsymbol{\delta}^w_{h^\clubsuit}(\mathfrak{L}_m\det[\boldsymbol{h}]), \tag{5.32}$$

that

$$\mathcal{R}_1(a) - \frac{1}{2}\boldsymbol{h}(a)\mathcal{R} - \lambda\boldsymbol{h}(a) = -\eta\boldsymbol{h}^{\clubsuit}T(a). \tag{5.33}$$

Multiplying Eq.(5.33) on both sides by $\boldsymbol{h}^{\dagger}\eta$ we get:

$$\boldsymbol{G}(a) - \lambda\boldsymbol{g}(a) = -T(a). \tag{5.34}$$

[5] Note that we are using geometrical units. The matter Lagrangian density is normally written as $\mathfrak{L}_m = -\kappa\mathfrak{L}'_m$, where $\kappa = -8\pi G$, where $G$ is Newton's gravitational constant.

# 6

# Gravitation Described by the Potentials $\mathfrak{g}^\alpha = \boldsymbol{h}^\dagger(\boldsymbol{\vartheta}^\alpha)$

ABSTRACT In this chapter, we will define in Section 6.1, the gravitational potentials $\mathfrak{g}^\alpha \in \sec \bigwedge^1 T^*U$, $\alpha = 0, 1, 2, 3$ as a result of a distortion by $\boldsymbol{h}^\dagger$of a 'cosmic lattice' characterizing the ground state of the Lorentz vacuum. Then, in Section 6.2, we will deduce from the Lagrangian for $\boldsymbol{h}^\clubsuit$(which includes a cosmological constant term) introduced in Chapter 5 the Lagrangian density for the potentials $\mathfrak{g}^\alpha$. The structure of this Lagrangian, is indeed, very interesting, it contains a term of the Yang-Mills type, a term (that for obvious reasons) we called a gauge fixing term and one which express the interaction of the 'vorticites' $\mathfrak{g}^\alpha \wedge d\mathfrak{g}_\alpha$ of the fields $\mathfrak{g}^\alpha$ with themselves. Section 6.3 deduces two possible expressions for the energy-momentum conservation law in our theory and Section 6.4 deduces a total angular momentum conservation law for out theory that includes the spin angular momentum of both the gravitational field and the matter fields. In Section 6.5, we will deduce the wave equations satisfied by the gravitational potentials in the effective Lorentzian spacetime generated by the gravitational field.

## 6.1 Definition of the Gravitational Potentials

Let $\{\mathrm{x}^\mu\}$be global coordinates in the Einstein-Lorentz Poincaré gauge for $M \simeq \mathbb{R}^4$and write, as above

$$\boldsymbol{\vartheta}^\mu = d\mathrm{x}^\mu. \tag{6.1}$$

Then

$$\boldsymbol{\eta} = \eta_{\alpha\beta}\boldsymbol{\vartheta}^\alpha \otimes \boldsymbol{\vartheta}^\beta. \tag{6.2}$$

V.V. Fernández and W.A. Rodrigues Jr., *Gravitation as a Plastic Distortion of the Lorentz Vacuum*,
Fundamental Theories of Physics 168, DOI 10.1007/978-3-642-13589-7_6,

We define next, the *gravitational potentials*[1]

$$\mathfrak{g}^\mu = \boldsymbol{h}^\dagger(\vartheta^\mu) \in \sec \bigwedge^1 T^*U, \tag{6.3}$$

and immediately have that

$$\mathfrak{g}^\alpha \underset{g^{-1}}{\cdot} \mathfrak{g}^\beta = \boldsymbol{h}^{-1}\eta \boldsymbol{h}^\clubsuit \mathfrak{g}^\alpha \cdot \mathfrak{g}^\beta = \eta \boldsymbol{h}^\clubsuit \mathfrak{g}^\alpha \cdot \boldsymbol{h}^\clubsuit \mathfrak{g}^\beta \tag{6.4}$$

$$= \boldsymbol{h}^\clubsuit \mathfrak{g}^\alpha \underset{\eta^{-1}}{\cdot} \boldsymbol{h}^\clubsuit \mathfrak{g}^\beta =$$

$$= \vartheta^\alpha \underset{\eta^{-1}}{\cdot} \vartheta^\beta = \vartheta^\alpha \underset{\eta}{\cdot} \vartheta^\beta = \eta^{\alpha\beta}, \tag{6.5}$$

i.e., the $\mathfrak{g}^\mu$ are $\boldsymbol{g}^{-1}$-orthonormal, which means that $\{\mathfrak{g}^\mu\}$ is a section of the $\boldsymbol{g}^{-1}$-orthonormal coframe bundle of $(M \simeq \mathbb{R}^4, \boldsymbol{g})$. Recalling now Eq.(4.117) which says that $R \equiv \overset{g}{R} = \mathcal{R}$, we can write the gravitational Lagrangian (Eq.(5.7)) as

$$\mathfrak{L}_{eh} := \frac{1}{2} R \det[\boldsymbol{h}], \tag{6.6}$$

and the Lagrangian density $\mathcal{L}_{eh} = \mathfrak{L}_{eh}\tau$ is then

$$\mathcal{L}_{eh} = \frac{1}{2}\{\boldsymbol{h}^\clubsuit(\partial_a \wedge \partial_b) \cdot \mathcal{R}_2(a \wedge b)\} \det[\boldsymbol{h}]\tau$$

$$= \frac{1}{2}\{\boldsymbol{h}^\clubsuit(\partial_a \wedge \partial_b) \cdot \eta \overset{\varkappa\eta}{\boldsymbol{R}}_2(a \wedge b)\}\tau_g, \tag{6.7}$$

where we have used the fact that $\tau_g = \det[\boldsymbol{h}]\tau$ and Eq.(4.112), which says that $\mathcal{R}_2(a \wedge b) = \eta \overset{\mu\eta}{\boldsymbol{R}}_2(a \wedge b)$. Now, we write the 1-form fields $a$,$b$ and the multiform derivatives $\partial_a, \partial_b$ as

$$a = a^\kappa \mathfrak{g}_\kappa,\ b = b^\kappa \mathfrak{g}_\kappa,$$

$$\partial_a = \mathfrak{g}^\kappa \frac{\partial}{\partial a^\kappa},\ \partial_b = \mathfrak{g}^\kappa \frac{\partial}{\partial b^\kappa}. \tag{6.8}$$

Then, using Eq.(6.8) in Eq.(6.7) and taking into account Eq.(4.111) which says that $\overset{g}{\boldsymbol{R}}(B) \equiv \overset{\lambda g}{\boldsymbol{R}}(B) = \underline{\boldsymbol{h}}^\dagger \overset{\varkappa\eta}{\boldsymbol{R}}_2(B)$ we get

$$2\mathcal{L}_{eh} = \{\boldsymbol{h}^\clubsuit(\mathfrak{g}^\kappa \wedge \mathfrak{g}^\iota) \cdot \eta \overset{\mu\eta}{\boldsymbol{R}}_2(\mathfrak{g}_\kappa \wedge \mathfrak{g}_\iota)\}\tau_g$$

$$= \{\boldsymbol{h}^\clubsuit(\mathfrak{g}^\kappa \wedge \mathfrak{g}^\iota) \cdot \eta \boldsymbol{h}^\clubsuit \boldsymbol{h}^\dagger \overset{\mu\eta}{\boldsymbol{R}}_2(\mathfrak{g}_\kappa \wedge \mathfrak{g}_\iota)\}\tau_g$$

$$= \{\boldsymbol{h}^{-1}\eta \boldsymbol{h}^\clubsuit(\mathfrak{g}^\kappa \wedge \mathfrak{g}^\iota) \cdot \boldsymbol{h}_2^\dagger \overset{\mu\eta}{\boldsymbol{R}}_2(\mathfrak{g}_\kappa \wedge \mathfrak{g}_\iota)\}$$

$$= \{\boldsymbol{h}^{-1}\eta \boldsymbol{h}^\clubsuit(\mathfrak{g}^\kappa \wedge \mathfrak{g}^\iota) \cdot \overset{g}{\boldsymbol{R}}_2(\mathfrak{g}_\kappa \wedge \mathfrak{g}_\iota)\}\tau_g$$

$$\{(\mathfrak{g}^\kappa \wedge \mathfrak{g}^\iota) \underset{g^{-1}}{\cdot} \overset{g}{\boldsymbol{R}}_2(\mathfrak{g}_\kappa \wedge \mathfrak{g}_\iota)\}\tau_g \tag{6.9}$$

---

[1]Note that, in the calculations of this section, we used the $\boldsymbol{g}$ reciprocal basis $\{\mathfrak{g}_\alpha\}$ of the basis $\{\mathfrak{g}^\mu\}$, i.e., $\mathfrak{g}^\mu \underset{g^{-1}}{\cdot} \mathfrak{g}_\alpha = \delta^\mu_\alpha$, i.e., $\mathfrak{g}_\alpha = \boldsymbol{h}^\dagger\eta(\vartheta_\alpha) = \eta_{\alpha\beta}\mathfrak{g}^\beta$.

We now recall that from Eq.(4.69), it is

$$\overset{g}{\boldsymbol{R}}_2(\mathfrak{g}_\kappa \wedge \mathfrak{g}_\iota) \equiv \mathcal{R}_{\kappa\iota} \tag{6.10}$$

where $\mathcal{R}_{\kappa\iota} : U \to \bigwedge^2 U$ are the representatives of the *Cartan curvature 2-form fields.* Then, recalling the definition of the Hodge star operator (Eq.(2.96)) we can write

$$\begin{aligned}\mathcal{L}_{eh} &= \frac{1}{2}\{(\mathfrak{g}^\kappa \wedge \mathfrak{g}^\iota) \underset{g^{-1}}{\cdot} \mathcal{R}_{\kappa\iota}\}\tau_g \\ &= \frac{1}{2}(\mathfrak{g}^\kappa \wedge \mathfrak{g}^\iota) \wedge \underset{g}{\star}\mathcal{R}_{\kappa\iota}\end{aligned} \tag{6.11}$$

Now, $\mathcal{L}_{eh}$ may be written, once we realize that the connection 1-forms can be written as

$$\omega^{\gamma\delta} = \frac{1}{2}\left[\mathfrak{g}^\delta \underset{g}{\lrcorner} d\mathfrak{g}^\gamma - \mathfrak{g}^\gamma \underset{g}{\lrcorner} d\mathfrak{g}^\delta + \mathfrak{g}^\gamma \underset{g}{\lrcorner}\left(\mathfrak{g}^\delta \underset{g}{\lrcorner} d\mathfrak{g}_\alpha\right)\mathfrak{g}^\alpha\right] \tag{6.12}$$

as

$$\mathcal{L}_{eh} = -\frac{1}{2}d\mathfrak{g}^\alpha \wedge \underset{g}{\star} d\mathfrak{g}_\alpha + \frac{1}{2}\underset{g}{\delta}\mathfrak{g}^\alpha \wedge \underset{g}{\star}\underset{g}{\delta}\mathfrak{g}_\alpha + \frac{1}{4}d\mathfrak{g}^\alpha \wedge \mathfrak{g}_\alpha \wedge \underset{g}{\star}(d\mathfrak{g}^\beta \wedge \mathfrak{g}_\beta) - d(\mathfrak{g}^\alpha \wedge \underset{g}{\star} d\mathfrak{g}_\beta) \tag{6.13}$$

where $d$ is the differential operator and $\underset{g}{\delta}$ is the Hodge coderivative operator given by a $r$-form field $A_r$ by (see e.g., [1] )

$$\underset{g}{\delta} A_r = (-1)^r \underset{g}{\star}^{-1} d \underset{g}{\star} A_r. \tag{6.14}$$

The proof of Eq.(6.13) is in Appendix D. Writing

$$\mathcal{L}_g = -\frac{1}{2}d\mathfrak{g}^\alpha \wedge \underset{g}{\star} d\mathfrak{g}_\alpha + \frac{1}{2}\underset{g}{\delta}\mathfrak{g}^\alpha \wedge \underset{g}{\star}\underset{g}{\delta}\mathfrak{g}_\alpha + \frac{1}{4}d\mathfrak{g}^\alpha \wedge \mathfrak{g}_\alpha \wedge \underset{g}{\star}(d\mathfrak{g}^\beta \wedge \mathfrak{g}_\beta) \tag{6.15}$$

and taking notice of an important identity [2]

$$-\frac{1}{2}d\mathfrak{g}^\alpha \wedge \underset{g}{\star} d\mathfrak{g}_\alpha + \frac{1}{2}\underset{g}{\delta}\mathfrak{g}^\alpha \wedge \underset{g}{\star}\underset{g}{\delta}\mathfrak{g}_\alpha = -\frac{1}{2}(d\mathfrak{g}^{\mathfrak{a}} \wedge \mathfrak{g}^\beta) \wedge \underset{g}{\star}(d\mathfrak{g}_\beta \wedge \mathfrak{g}_\alpha), \tag{6.16}$$

we can write Eq.(6.15) as

$$\mathcal{L}_g = -\frac{1}{2}(d\mathfrak{g}^\alpha \wedge \mathfrak{g}^\beta) \wedge \underset{g}{\star}(d\mathfrak{g}_\beta \wedge \mathfrak{g}_\alpha) + \frac{1}{4}d\mathfrak{g}^\alpha \wedge \mathfrak{g}_\alpha \wedge \underset{g}{\star}(d\mathfrak{g}^\beta \wedge \mathfrak{g}_\beta) \tag{6.17}$$

Using Eq.(2.94) and Eq.(2.101), we can write:

$$\begin{aligned}(d\mathfrak{g}^\alpha \wedge \mathfrak{g}^\beta) \wedge \underset{g}{\star}(d\mathfrak{g}_\alpha \wedge \mathfrak{g}_\beta) &= (d\mathfrak{g}^\alpha \wedge \mathfrak{g}^\beta) \wedge \underline{\boldsymbol{h}}^\dagger \underset{\eta}{\star} \underline{\boldsymbol{h}}^\clubsuit(d\mathfrak{g}_\alpha \wedge \mathfrak{g}_\beta) \\ &= \underline{\boldsymbol{h}}^\dagger[\underline{\boldsymbol{h}}^\clubsuit d\mathfrak{g}^\alpha \wedge \boldsymbol{h}^\clubsuit\mathfrak{g}^\beta \wedge \underset{\eta}{\star}\underline{\boldsymbol{h}}^\clubsuit d\mathfrak{g}_\alpha \wedge \boldsymbol{h}^\clubsuit\mathfrak{g}_\beta].\end{aligned} \tag{6.18}$$

$$\tag{6.19}$$

$$(d\mathfrak{g}^\alpha \wedge \mathfrak{g}_\alpha) \wedge \underset{g}{\star}(d\mathfrak{g}^\beta \wedge \mathfrak{g}_\beta) = \underline{\boldsymbol{h}}^\dagger[\underline{\boldsymbol{h}}^\clubsuit d\mathfrak{g}^\alpha \wedge \boldsymbol{h}^\clubsuit\mathfrak{g}_\alpha \wedge \underset{\eta}{\star}\underline{\boldsymbol{h}}^\clubsuit d\mathfrak{g}^\beta \wedge \boldsymbol{h}^\clubsuit\mathfrak{g}_\beta)]. \tag{6.20}$$

and write $2\mathcal{L}_h = \underline{2\boldsymbol{h}^\clubsuit}\mathcal{L}_g$ as

$$2\mathcal{L}_h = -\left(\underline{\boldsymbol{h}}^\clubsuit(\mathfrak{g}^\alpha \wedge \mathfrak{g}^\beta) \underset{\eta^{-1}}{\cdot} \underline{\boldsymbol{h}}^\clubsuit(d\mathfrak{g}_\alpha \wedge \mathfrak{g}_\beta) - \frac{1}{2}\underline{\boldsymbol{h}}^\clubsuit(d\mathfrak{g}^\alpha \wedge \mathfrak{g}_\alpha) \underset{\eta^{-1}}{\cdot} \underline{\boldsymbol{h}}^\clubsuit(d\mathfrak{g}^\beta \wedge \mathfrak{g}_\beta)\right)\tau_\eta \tag{6.21}$$

**Remark 6.1** A careful inspection of the Lagrangian density given by Eq.(6.15) reveals its remarkable structure. Indeed, the first term is of the Yang-Mills kind and the second term may be called a gauge fixing term, since, e.g., $\underset{g}{\delta}\mathfrak{g}^\alpha = 0$ is analogous of the *Lorenz gauge* in electromagnetic theory. Finally, the third term is an auto-interaction term, describing the coupling of the *vorticities*[2] $(d\mathfrak{g}^\alpha \wedge \mathfrak{g}_\alpha)$ of the fields $\mathfrak{g}^\alpha$. One way of saying that is: the distortion field $\boldsymbol{h}$ puts the Lorentz vacuum medium in motion. That motion is described by $\mathfrak{g}^\alpha = \boldsymbol{h}^\dagger(\vartheta^\alpha)$, saying that the deformed cosmic lattice $\boldsymbol{h}^\dagger(\vartheta^\alpha)$ is in motion relative to the ground state cosmic lattice which is defined (modulo a global Lorentz transformation) by $\{\vartheta^\alpha\}$.

**Remark 6.2** In the gravitational theory, with the Lagrangian density given by Eq.(6.15), no mention of *any* connection on the world manifold appears. But, if some one wants to give a geometrical model interpretation for such a theory, of course, the simple one is to say that the gravitational field generates a teleparallel geometry with a metric compatible connection defined on the world manifold by[3] $\nabla^-_{\mathfrak{e}_\alpha}\mathfrak{g}^\beta = 0$, $\nabla^-\boldsymbol{g} = 0$. For that connection, the Riemann curvature tensor is null whereas the torsion 2-form fields are given by $\Theta^\alpha = d\mathfrak{g}^\alpha$ (i.e., the gravitational field is torsion in this model). Only on second thoughts, someone would think to introduce a Levi-Civita connection $D$ on the world manifold such that $D^-_{\mathfrak{e}_\alpha}\mathfrak{g}^\beta = -L^\beta_{\alpha\gamma}\mathfrak{g}^\gamma$, $D^-\boldsymbol{g} = 0$, for which the torsion 2-forms $\Theta^\alpha = d\mathfrak{g}^\alpha + \omega^\alpha_\beta \wedge \mathfrak{g}^\beta = 0$ and the curvature 2-forms $\mathcal{R}^\alpha_\beta = d\omega^\alpha_\beta + \omega^\alpha_\beta \wedge \omega^\alpha_\beta \neq 0$. However, since history did not follow that path, it was the geometry associated with the Levi-Civita connection that was first discovered by Einstein and Grossmann. More on this issue may be found in [4] .

**Remark 6.3** Eq.(6.21) expresses $\mathcal{L}_h$ as a functional of $\boldsymbol{h}^\clubsuit$ and uses only objects belonging to the Minkowski spacetime structure. However, to perform the variation of $\mathcal{L}_h$, in order to determine the equations of motion for the extensor field $\boldsymbol{h}^\clubsuit$ is an almost impracticable task, which may lead one to appreciate the tricks used in Chapter 5 to derive the equation of motion for that extensor field $\boldsymbol{h}$. The derivation of the equations of motion directly for the potential fields $\mathfrak{g}^\mathbf{a}$ is an easier (but still involved) task and is given in Appendix E.

## 6.2 Lagrangian Density for the Massive Gravitational Field Plus the Matter Fields

In this section, we study the interaction of the gravitational potentials $\mathfrak{g}^\mathbf{a}$ with the matter fields described by a Lagrangian density $\mathcal{L}_m$. Moreover, as before in Section 5.1.2, we add a term corresponding to the Lorentz vacuum energy density, the one given in terms of the cosmological constant, which we prefer *here* to write as a 'graviton mass' term, i.e., we have:

$$\mathcal{L}' = \mathcal{L}_{eh} + \frac{1}{2}m^2\mathfrak{g}_\alpha \wedge \underset{g}{\star}\mathfrak{g}^\alpha + \mathcal{L}_m \tag{6.22}$$

---

[2]The name vorticity has been used here for the following reason. In *GRT*, as well known (see, e.g., [1,3]) any given *reference frame* existing in $(M, \boldsymbol{g}, D, \tau_{\boldsymbol{g}}$ is represented by a given timelike vector field $Z$ (pointing to the future). The physically equivalent 1-form $\alpha = \boldsymbol{g}(\ ,Z)$ has the unique decomposition $D\alpha = \mathbf{a}\otimes\alpha + \omega + \sigma + \frac{1}{3}\mathfrak{E}\mathbf{h}$,where $\mathbf{h} \in \sec T^0_2 M$ is the projection tensor $\mathbf{h} = \mathbf{g} - \alpha\otimes\alpha$, $\mathbf{a}$ *acceleration* of $Z$, $\omega$ is the *rotation* tensor (or *vortex*) of $Z$, $\sigma$ is the *shear* of $Z$ and $\mathfrak{E}$ is the *expansion ratio* of $Z$. We can show that the non null components of $\omega$ are the same as the ones in $\alpha \wedge d\alpha$.

[3]The $\{\mathfrak{e}_\alpha\}$ is the dual basis of $\{\mathfrak{g}^\alpha\}$.

or equivalently, since $\mathcal{L}_{eh}$ and $\mathcal{L}_g$ differs from an exact differential, we shall use

$$\begin{aligned}\mathcal{L} &= \mathcal{L}_g + \frac{1}{2}m^2\mathfrak{g}_\alpha \wedge \underset{g}{\star}\mathfrak{g}^\alpha + \mathcal{L}_m \\ &= \mathcal{L}'_g + \mathcal{L}_m.\end{aligned} \tag{6.23}$$

The resulting equations of motion (whose derivation for completeness is given in the Appendix E) may be written putting

$$\star t^\alpha = \frac{\partial \mathcal{L}'_g}{\partial \mathfrak{g}_\alpha}, \ \underset{g}{\star}\mathcal{S}^\alpha = \frac{\partial \mathcal{L}'_g}{\partial d\mathfrak{g}_\alpha}, \ -\underset{g}{\star}\mathcal{T}^\alpha = \underset{g}{\star}\mathcal{T}^\alpha := -\frac{\partial \mathcal{L}_m}{\partial \mathfrak{g}_\alpha}, \tag{6.24}$$

as

$$-\underset{g}{\star}G^\alpha = d\underset{g}{\star}\mathcal{S}^\alpha + \underset{g}{\star}\mathfrak{t}^\alpha = -\underset{g}{\star}\mathcal{T}^\alpha, \tag{6.25}$$

where $\underset{g}{\star}G^\alpha \in \sec\bigwedge^3 T^*U$ are the Einstein 3-form fields while, the $\underset{g}{\star}\mathfrak{t}^\kappa \in \sec\bigwedge^3 T^*U$ and the $\underset{g}{\star}\mathcal{S}^\alpha \in \sec\bigwedge^2 T^*U$ are given by (recall Eq.(E.30) and Eq.(E.31) of Appendix E):

$$\underset{g}{\star}\mathfrak{t}^\kappa = \frac{\partial \mathcal{L}'_g}{\partial \mathfrak{g}_\kappa} = \frac{\partial \mathcal{L}_g}{\partial \mathfrak{g}_\kappa} + m^2\mathfrak{g}^\kappa = \mathfrak{g}^\kappa \underset{g}{\lrcorner}\mathcal{L}_g - (\mathfrak{g}^\kappa \underset{g}{\lrcorner} d\mathfrak{g}_\alpha) \wedge \underset{g}{\star}\mathcal{S}^\alpha + m^2\mathfrak{g}^\kappa \tag{6.26}$$

$$\underset{g}{\star}\mathcal{S}^\kappa = \frac{\partial \mathcal{L}_g}{\partial d\mathfrak{g}_\kappa} = -\mathfrak{g}^\alpha \wedge \underset{g}{\star}(d\mathfrak{g}_\alpha \wedge \mathfrak{g}^\kappa) + \frac{1}{2}\mathfrak{g}^\kappa \wedge \underset{g}{\star}(d\mathfrak{g}^\alpha \wedge \mathfrak{g}_\alpha). \tag{6.27}$$

We write, moreover,

$$\underset{g}{\star}\mathcal{S}^\kappa = \underset{g}{\star}d\mathfrak{g}^\kappa + \underset{g}{\star}\mathfrak{K}^\kappa, \tag{6.28}$$

$$\underset{g}{\star}\mathfrak{K}^\kappa = -(\mathfrak{g}^\kappa \underset{g}{\lrcorner}\underset{g}{\star}\mathfrak{g}^\alpha) \wedge \underset{g}{\star}d\underset{g}{\star}\mathfrak{g}_\alpha + \frac{1}{2}\mathfrak{g}^\kappa \wedge \underset{g}{\star}(d\mathfrak{g}^\alpha \wedge \mathfrak{g}_\alpha) \tag{6.29}$$

and insert this result in Eq.(6.25), obtaining:

$$d\underset{g}{\star}d\mathfrak{g}^\kappa + m^2\underset{g}{\star}\mathfrak{g}^\kappa = -\underset{g}{\star}\left(\mathfrak{t}^\kappa + \mathcal{T}^\kappa + \underset{g}{\star}^{-1}d\underset{g}{\star}\mathfrak{K}^\kappa\right) \tag{6.30}$$

Applying the operator $\underset{g}{\star}^{-1}$ to both sides of that equation, we get

$$-\underset{g}{\delta}d\mathfrak{g}^\kappa + m^2\mathfrak{g}^\kappa = -\left(t^\kappa + \mathcal{T}^\kappa + \underset{g}{\delta}\mathfrak{K}^\kappa\right) \tag{6.31}$$

Calling

$$\mathfrak{F}^\kappa := d\mathfrak{g}^\kappa$$

we have the following *Maxwell like* equations for the gravitational fields $\mathfrak{F}^\kappa$:

$$d\mathfrak{F}^\kappa = 0, \tag{6.32a}$$

$$\underset{g}{\delta}\mathfrak{F}^\kappa = \left(t^\kappa + \mathfrak{T}^\kappa + m^2\mathfrak{g}^\kappa\right), \tag{6.32b}$$

$$\mathfrak{T}^\kappa = \mathcal{T}^\kappa + \underset{g}{\delta}\mathfrak{K}^\kappa \tag{6.32c}$$

However, note that the above system of differential equations is non linear in the gravitational potentials $\mathfrak{g}^\kappa$.

Now we may define the total energy-momentum of the system consisting of the matter plus the gravitational field either as or $\mathbf{P} = P_\kappa \vartheta^\kappa$

$$P_\kappa := -\int_U \underset{g}{\star}(\mathfrak{t}_\kappa + \mathfrak{T}^\kappa + m^2 \mathfrak{g}_\kappa) = \int_{\partial U} \underset{g}{\star} d\mathfrak{g}_\kappa. \tag{6.33}$$

or $\mathbf{P}' = P'_\kappa \vartheta^\kappa$

$$P'_\kappa := -\int_U \underset{g}{\star}(\mathfrak{t}_\kappa + \mathcal{T}_\kappa + m^2 \mathfrak{g}_\kappa) = \int_{\partial U} \underset{g}{\star} \mathcal{S}_\kappa, \tag{6.34}$$

## 6.3 Energy-Momentum Conservation Law

The Maxwell like formulation (Eq. (6.32)) or Eq.(6.25)) immediately imply two possible energy-momentum conservation laws in our theory (i.e., both the $P_\kappa$ as well as the $P'_\kappa$ are conserved), but in order to avoid any misunderstanding some remarks are necessary.

**Remark 6.4** The values of $P_\kappa$ (Eq.(6.33)) as well as $P'_\kappa$ (Eq.(6.34)) are, of course, independent of the *coordinate system* used to calculate them. This is also the case in *GRT*, where identical equations hold for models with global **g**-orthonormal cotetrad fields. However, in *GRT* expressions anlagous to $\mathbf{P} = P_\kappa \vartheta^\kappa$ or $\mathbf{P}' = P'_\kappa \vartheta^\kappa$ can be given meaning only for asymptotically flat spacetimes. Also, it is obvious that, in our theory, as well as in *GRT* both $P_\kappa$ or $P'_\kappa$ depend on the choice of the cotetrad field $\{\mathfrak{g}^\alpha\}$. The difference is that in our theory, the $\{\mathfrak{g}^\alpha\}$, the gravitational potentials (defined by the extensor field $\boldsymbol{h}$ modulus an arbitrary local Lorentz rotation $\Lambda$ which is hidden in the definition of $\boldsymbol{g} = \boldsymbol{h}^\dagger \eta \boldsymbol{h} = \boldsymbol{h}^\dagger \Lambda^\dagger \eta \Lambda \boldsymbol{h})$ are by chance a section of the **g**-orthonormal frame bundle of an effective Lorentzian spacetime (with the $\underset{g}{\star} t_\kappa$ being thus gauge dependent objects) whereas in *GRT* the set $\{\mathfrak{g}^\alpha\}$ does not have any interpretation different from the one of being a section of the **g**-orthonormal frame bundle of the Lorentzian spacetime representing a particular gravitational field. Moreover in *GRT* we can write equations similar to Eq.(6.26) also for coordinate basis of the linear frame bundle of the Lorentzian spacetime, thus obtaining different 'energy-mometum pseudo-tensors', for which the *values* of the integral analogous to Eq.(6.34) depend on the coordinate system used for their calculations, something that is clearly a nonsense [5,6].

**Remark 6.5** The condition of possessing a global **g**-orthonormal tetrad field is, of course, not satisfied by a general Lorentzian structure[4] $(M, \mathbf{g})$ and, if we want *GRT* to admits a total energy-momentum conservation law for the system consisting of the matter and gravitational fields, such a condition must result as consequence of other *extra* condition imposed on the spacetime structure, a fact that we already mentioned in Chapter 1.

**Remark 6.6** Another very well known fact concerning the $t^\alpha$ is that their components are not symmetric. Then, for example in [8] it is claimed that in order to obtain an angular momentum conservation law, it is necessary to use superpotentials different from the $\mathcal{S}^\alpha$ which produce a symmetric $t^\alpha$. The choice made, e.g., in [8] is then to use the so-called Landau-Lifschitz pseudo-energy momentum tensor [9], which, as is well known, is a symmetric object[5]. However, we will show

[4] Such a condition, as is well known [1,6] is a necessary one for the existence of spinor fields.

[5] For more details on those issues see [6].

next, that with the original non symmetric $t^{\alpha}$, we can get a total angular momentum conservation law which includes the spin density of the gravitational field.

## 6.4 Angular Momentum Conservation Law

Consider the global chart for $\mathbb{R}^4$ with coordinates in the Einstein-Lorentz-Poincaré gauge $\{\mathrm{x}^{\mu}\}$ and as previously put $\boldsymbol{\vartheta}^{\mu} = d\mathrm{x}^{\mu}$. Calling

$$\underset{g}{\star}\mathbf{T}^{\alpha} := \underset{g}{\star}(t^{\kappa} + \mathfrak{T}^{\kappa} + +m^2\mathfrak{g}^{\kappa}) \tag{6.35}$$

the (non symmetric) *total* energy momentum 3-forms of matter plus the gravitational field, we define the *chart dependent* density of *total orbital angular momentum* of the matter as the 3-form fields

$$\underset{g}{\star}\mathbf{L}_m^{\alpha\beta} := \underline{\boldsymbol{h}}^{\dagger}\{\boldsymbol{h}^{\clubsuit}(\mathrm{x}^{\alpha}) \wedge \underset{\eta}{\star}\boldsymbol{h}^{\clubsuit}\mathbf{T}^{\beta} - \boldsymbol{h}^{\clubsuit}(\mathrm{x}^{\beta}) \wedge \underset{\eta}{\star}\boldsymbol{h}^{\clubsuit}\mathbf{T}^{\alpha}\}, \tag{6.36}$$

which can be written as

$$\underset{g}{\star}\mathbf{L}_m^{\alpha\beta} = \mathrm{x}^{\alpha} \underset{g}{\star} \mathbf{T}^{\beta} - \mathrm{x}^{\beta} \underset{g}{\star} \mathbf{T}^{\alpha}. \tag{6.37}$$

Of course,

$$d \underset{g}{\star} \mathbf{L}_m^{\alpha\beta} = \boldsymbol{\vartheta}^{\alpha} \wedge \underset{g}{\star}\mathbf{T}^{\beta} - \boldsymbol{\vartheta}^{\beta} \wedge \underset{g}{\star}\mathbf{T}^{\alpha} \neq 0. \tag{6.38}$$

However, let us define the density (3-forms) of *orbital* angular momentum of the gravitational field by

$$\underset{g}{\star}\mathbf{L}_g^{\alpha\beta} := \underline{\boldsymbol{h}}^{\dagger}\{\mathfrak{g}^{\alpha} \wedge \underset{\eta}{\star}\boldsymbol{h}^{\clubsuit}\mathfrak{F}^{\beta} - \mathfrak{g}^{\beta} \wedge \underset{\eta}{\star}\boldsymbol{h}^{\clubsuit}\mathfrak{F}^{\alpha}\}, \tag{6.39}$$

which can be written as

$$\underset{g}{\star}\mathbf{L}_g^{\alpha\beta} = \boldsymbol{\vartheta}^{\alpha} \wedge \underset{g}{\star}\mathfrak{F}^{\beta} - \boldsymbol{\vartheta}^{\beta} \wedge \underset{g}{\star}\mathfrak{F}^{\alpha}, \tag{6.40}$$

and the density of *total orbital angular momentum* of the matter plus the gravitational field by

$$\underset{g}{\star}\mathbf{L}_t^{\alpha\beta} = \underset{g}{\star}\mathbf{L}_m^{\alpha\beta} + \underset{g}{\star}\mathbf{L}_g^{\alpha\beta}. \tag{6.41}$$

Then, we immediately get that

$$\begin{aligned} d \underset{g}{\star} \mathbf{L}_t^{\alpha\beta} &= \boldsymbol{\vartheta}^{\alpha} \wedge \underset{g}{\star}\mathbf{T}^{\beta} - \boldsymbol{\vartheta}^{\beta} \wedge \underset{g}{\star}\mathbf{T}^{\alpha} - \boldsymbol{\vartheta}^{\alpha} \wedge d \underset{g}{\star} \mathfrak{F}^{\beta} + \boldsymbol{\vartheta}^{\beta} \wedge d \underset{g}{\star} \mathfrak{F}^{\alpha} \\ &= \boldsymbol{\vartheta}^{\alpha} \wedge \underset{g}{\star}\mathbf{T}^{\beta} - \boldsymbol{\vartheta}^{\beta} \wedge \underset{g}{\star}\mathbf{T}^{\alpha} - \boldsymbol{\vartheta}^{\alpha} \wedge \underset{g}{\star}\mathbf{T}^{\beta} + \boldsymbol{\vartheta}^{\beta} \wedge \underset{g}{\star}\mathbf{T}^{\alpha} \\ &= 0. \end{aligned} \tag{6.42}$$

However, Eq.(6.42) does not contain, as yet the spin angular momentum of matter and the spin angular momentum of the gravitational field and thus cannot be considered as a satisfactory equation for the conservation of total angular momentum. To go on, we proceed as follows. First we write the total Lagrangian density for the gravitational field as

$$\mathcal{L} = \mathcal{L}_{eh} + \mathcal{L}_m, \tag{6.43}$$

where $\mathcal{L}_{eh}$ is defined by Eq.(6.11), i.e.,

$$\mathcal{L}_{eh} = \frac{1}{2}\mathcal{R}_{\kappa\iota} \wedge \underset{g}{\star}(\mathfrak{g}^\kappa \wedge \mathfrak{g}^\iota), \tag{6.44}$$

but where now we suppose that $\mathcal{L}_{eh}$ is a functional of the gravitational potentials $\mathfrak{g}^\kappa$ and the connection 1-form fields $\omega_{\kappa\iota}$ appearing in $\mathcal{R}_{\kappa\iota}$ which we take as independent variables to start. Moreover, we suppose that $\mathcal{L}_m$ depends on the field variables $\phi^A$ (which in general are indexed form fields[6]) and their exterior covariant derivatives, i.e., it is a functional of the kind $\mathcal{L}_m = \mathcal{L}_m(\phi^A, d\phi^A, \omega_{\kappa\iota})$.

If we make the variation of the action $\int_U (\mathcal{L}_{eh} + \mathcal{L}_m)$ with respect to $\omega_{\kappa\iota}$ we get

$$\begin{aligned} &\delta \int_U (\mathcal{L}_{eh} + \mathcal{L}_m) \\ &= \int_U (\delta\omega_{\kappa\iota} \wedge \frac{\partial \mathcal{L}_{eh}}{\partial \omega_{\kappa\iota}} + \delta d\omega_{\kappa\iota} \wedge \frac{\partial \mathcal{L}_{eh}}{\partial d\omega_{\kappa\iota}} + \delta\omega_{\kappa\iota} \wedge \frac{\partial \mathcal{L}_m}{\partial \omega_{\kappa\iota}}) \\ &= \int_U (\delta\omega_{\kappa\iota} \wedge \left[\frac{\partial \mathcal{L}_{eh}}{\partial \omega_{\kappa\iota}} + d\left(\frac{\partial \mathcal{L}_{eh}}{\partial d\omega_{\kappa\iota}}\right) + \frac{\partial \mathcal{L}_m}{\partial \omega_{\kappa\iota}}\right] + d\left(\delta\omega_{\kappa\iota} \wedge \frac{\partial \mathcal{L}_{eh}}{\partial d\omega_{\kappa\iota}}\right), \end{aligned} \tag{6.45}$$

from where we obtain from the principle of stationary action and with the usual hypothesis that the variations vanishes on the boundary $\partial U$ of $U$ that

$$\frac{\partial \mathcal{L}_{eh}}{\partial \omega_{\kappa\iota}} + d\left(\frac{\partial \mathcal{L}_{eh}}{\partial d\omega_{\kappa\iota}}\right) + \frac{\partial \mathcal{L}_m}{\partial \omega_{\kappa\iota}} = 0 \tag{6.46}$$

In metric affine theories the spin [10,11] (density) angular momentum of matter is defined by

$$\underset{g}{\star}\mathbf{S}_m^{\kappa\iota} = \frac{\partial \mathcal{L}_m}{\partial \omega_{\kappa\iota}}. \tag{6.47}$$

and we will accept it as a good one[7]. Now,

$$\begin{aligned} \frac{\partial \mathcal{L}_{eh}}{\partial \omega_{\kappa\iota}} &= \frac{1}{2}\left(\omega_\alpha^\kappa \wedge \underset{g}{\star}(\mathfrak{g}^\alpha \wedge \mathfrak{g}^\iota) + \omega_\alpha^\iota \wedge \underset{g}{\star}(\mathfrak{g}^\kappa \wedge \mathfrak{g}^\alpha)\right), \\ d\left(\frac{\partial \mathcal{L}_{eh}}{\partial d\omega_{\kappa\iota}}\right) &= \frac{1}{2}\underset{g}{\star}(\mathfrak{g}^\kappa \wedge \mathfrak{g}^\iota). \end{aligned} \tag{6.48}$$

Thus, defining

$$\underset{g}{\star}\mathbf{S}_g^{\kappa\iota} := \omega_\alpha^\kappa \wedge \underset{g}{\star}(\mathfrak{g}^\alpha \wedge \mathfrak{g}^\iota) + \omega_\alpha^\iota \wedge \underset{g}{\star}(\mathfrak{g}^\kappa \wedge \mathfrak{g}^\alpha) \tag{6.49}$$

as the spin (density) angular momentum of the gravitational field, we get

$$d\underset{g}{\star}(\mathfrak{g}^\kappa \wedge \mathfrak{g}^{\mathrm{l}}) = -2\underset{g}{\star}\left(\mathbf{S}_m^{\kappa\iota} + \mathbf{S}_g^{\kappa\iota}\right), \tag{6.50}$$

---

[6]Even spinor fields may be represented in he Clifford bundle formalism as some equivalence classes of non homogenous form fields. See, e.g., [1] for details.

[7]In particular, in [11], it is shown that this agrees with the spin density, which has as source the antisymmetric part of the canonical energy-momentum tensor of the *matter* field.

and thus[8]

$$d \underset{g}{\star} (\mathbf{S}_m^{\kappa\iota} + \mathbf{S}_g^{\kappa\iota}) = 0. \tag{6.51}$$

We now define the total angular momentum of the matter field plus the gravitational field as

$$\underset{g}{\star}\mathbf{J}^{\alpha\beta} := \underset{g}{\star}(\mathbf{L}_m^{\alpha\beta} + \mathbf{S}_m^{\kappa\iota}) + \underset{g}{\star}(\mathbf{L}_g^{\alpha\beta} + \mathbf{S}_g^{\kappa\iota}), \tag{6.52}$$

and due Eq.(6.42) and Eq.(6.51) we have

$$d \underset{g}{\star} \mathbf{J}^{\alpha\beta} = 0. \tag{6.53}$$

Eq.(6.53) expresses the law of conservation of total angular momentum in our theory, and surprisingly, it seems that the orbital and spin angular momentum are separately conserved.

## 6.5 Wave Equations for the $\mathfrak{g}^\kappa$

The wave equation for the gravitational potential $\mathfrak{g}^\kappa$ can be written easily in the *effective* Lorentzian spacetime structure $(M, \boldsymbol{g}, D, \tau_{\boldsymbol{g}}, \uparrow)$ already introduced. Indeed, to get that wave equation, we add the term $d\underset{g}{\delta}\mathfrak{g}^\kappa$ to both members of Eq.(6.31) getting

$$-\underset{g}{\delta}d\mathfrak{g}^\kappa - d\underset{g}{\delta}\mathfrak{g}^\kappa + m^2\mathfrak{g}^\kappa = -\left(\mathfrak{t}^\kappa + \mathcal{T}^\kappa + \underset{g}{\delta}\mathfrak{R}^\kappa + d\underset{g}{\delta}\mathfrak{g}^\kappa\right). \tag{6.54}$$

We now recall the definition of the Hodge D'Alembertian, which in the Clifford bundle formalism is the square of the Dirac operator ($\boldsymbol{\partial} := \mathfrak{g}^\kappa \underset{g}{} D^-_{\mathfrak{g}_\kappa} = d - \underset{g}{\delta}$), acting on multiform fields [1], i.e., we can write

$$\mathring{\Diamond}\mathfrak{g}^\kappa := \boldsymbol{\partial}^2\mathfrak{g}^\kappa = (-\underset{g}{\delta}d - d\underset{g}{\delta})\mathfrak{g}^\kappa. \tag{6.55}$$

We recall, moreover, the following nontrivial decomposition [1] of $\boldsymbol{\partial}^2$,

$$\boldsymbol{\partial}^2\mathfrak{g}^\kappa = \boldsymbol{\partial} \underset{g}{\cdot} \boldsymbol{\partial}\mathfrak{g}^\kappa + \boldsymbol{\partial} \wedge \boldsymbol{\partial}\mathfrak{g}^\kappa, \tag{6.56}$$

where $\square := \boldsymbol{\partial} \underset{g}{\cdot} \boldsymbol{\partial}$ is the *covariant D'Alembertian* and $\boldsymbol{\partial} \wedge \boldsymbol{\partial}$ is the *Ricci operator associated to the Levi-Civita connection of* $\boldsymbol{g}$. If

$$D^-_{\mathfrak{g}_\alpha}\mathfrak{g}^\beta = -L^\beta_{\alpha\rho}\mathfrak{g}_\rho, \tag{6.57}$$

then

$$\begin{array}{lrcl} \text{(a)} & \boldsymbol{\partial} \underset{g}{\cdot} \boldsymbol{\partial} & = & \eta^{\alpha\beta}(D^-_{\mathfrak{g}_\alpha}D^-_{\mathfrak{g}_\beta} - L^\rho_{\alpha\beta}D^-_{\mathfrak{g}_\rho}) \\ \text{(b)} & \boldsymbol{\partial} \wedge \boldsymbol{\partial} & = & \mathfrak{g}^\alpha \wedge \mathfrak{g}^\beta(D^-_{\mathfrak{g}_\alpha}D^-_{\mathfrak{g}_\beta} - L^\rho_{\alpha\beta}D^-_{\mathfrak{g}_\rho}), \end{array} \tag{6.58}$$

and a simple computation shows that

$$\boldsymbol{\partial} \wedge \boldsymbol{\partial}\mathfrak{g}^\kappa = \mathcal{R}^\kappa, \tag{6.59}$$

[8] An equation analogous to this one was found originally by Bramsom [12].

where $\mathcal{R}^{\kappa} : U \to \bigwedge^1 U$ are the Ricci 1-form fields, given by:

$$\mathcal{R}^{\kappa} = R^{\kappa}_{\iota}\mathfrak{g}^{\iota}, \tag{6.60}$$

where $R^{\kappa}_{\iota}$ are the components of the Ricci tensor in the basis defined by $\{\mathfrak{g}^{\alpha}\}$. Then, we have the following wave equations for the gravitational potentials:

$$\Box\mathfrak{g}^{\kappa} + m^2\mathfrak{g}^{\kappa} = -(\mathcal{T}^{\kappa} + \mathfrak{t}^{\kappa} + \underset{g}{\delta}\mathfrak{K}^{\kappa} + d\underset{g}{\delta}\mathfrak{g}^{\kappa}). \tag{6.61}$$

## References


[1] Rodrigues, W. A. Jr., and Capelas de Oliveira, E., T*he Many Faces of Maxwell, Dirac and Einstein Equations. A Clifford Bundle Approach.* Lecture Notes in Physics **722**, Springer, Heidelberg, 2007.

[2] Rodrigues, W. A. Jr., and Souza, Q. A. G., An Ambiguous Statement Called 'Tetrad Postulate' and the Correct Field Equations Satisfied by the Tetrad Fields, *Int.J.Mod.Phys D* **14**, 2095-2150 (2005). [arXiv:math-ph/0411085]

[3] Sachs, R. K., and Wu, H., *General Relativity for Mathematicians*, Springer-Verlag, New York, 1977.

[4] Notte-Cuello, E. A., da Rocha, R., and Rodrigues, W. A. Jr., *Some Thoughts on Geometries and on the Nature of the Gravitational Field.* [axXiv:0907.2424]

[5] Bohzkov, Y., and Rodrigues, W. A. Jr., Mass and Energy in General Relativity, *Gen. Rel. Grav.* **27**, 813- 819 (1995).

[5] Notte-Cuello, E. A., and Rodrigues, W. A. Jr., Freud's Identity of Differential Geometry, the Einstein-Hilbert Equations and the Vexatious Problem of the Energy-Momentum Conservation in GR, *Adv. Appl. Cliford Algebras* **19,** 113-145 (2009). [arXiv:0801.2559]

[6] Geroch, R., Spinor Structure of Space-Times in General Relativity I, *J. Math. Phys.* **9**, 1739-1744 (1988).

[7] Thirring, W., *A Course in Mathematical Physics*, vol. **2**., Springer-Verlag, New York, 1979.

[8] Landau, L.D. and Lifshitz, E. M., *The Classical Theory of Fields* (fourth revised english edition), Pergamon Presss, New York, 1975.

[9] Benn, I. M., Conservation Laws in Arbitrary Space-times, *Ann. Inst. H. Poincaré*, **XXXVII**, 67-91 (1982).

[10] Hehl, F. W., von der Heyde, P., Kerlick, G. D., and Nester, J. M., General Relativity with Spin and Torsion: Foundations and Prospects, *Rev. Mod. Phys.* **48**, 393-416 (1976).

[11] Bramson, B. D., Relativistic Angular Momentum for Asymptotically Flat Einstein-Maxwell Manifolfds, *Proc. R. Soc. London Ser. A* **341**, 463-469 (1975).

# 7

# Hamiltonian Formalism

ABSTRACT In this chapter, we will investigate the Hamiltonian formalism for our gravitational theory, when expressed in terms of the gravitational potentials. So, in Section 7.1, after briefly recalling the concept of a Legendre transformation we will obtain several different equivalent expressions for the Hamiltonian 3-form density $\mathcal{H}$ of our theory. In Section 7.2,. we briefly recall the concept of quasi-local energy (which is of importance in *GRT*) and show how it applies naturally to our theory. Moreover, we will show that, for a special choice of a timelike vector field defining the 'flow of time' in the Hamiltonian formalism, the energy of a gravitational configuration yields an identical result with one calculated with the approach of Section 6.4 and given by Eq.(6.34). In Section 7.4, we will deduce Hamilton's equations of motion. Finally, in section 7.5, we will make contact between our Hamiltonian formalism and the celebrated *ADM* formalism of *GRT*, and show that the *ADM* energy coincides again with the one given by Eq.(6.34), for a special choice of a timelike vector field defining the 'flow of time'.

## 7.1 The Hamiltonian 3-form Density $\mathcal{H}$

We start with the Lagrangian density[1] given by Eq.(6.15), i.e.,

$$\mathcal{L}_g(\mathfrak{g}^\alpha, d\mathfrak{g}^\alpha) = -\frac{1}{2}d\mathfrak{g}^\alpha \wedge \underset{g}{\star} d\mathfrak{g}_\alpha + \frac{1}{2}\underset{g}{\delta}\mathfrak{g}^\alpha \wedge \underset{g}{\star}\underset{g}{\delta}\mathfrak{g}_\alpha + \frac{1}{4}d\mathfrak{g}^\alpha \wedge \mathfrak{g}_\alpha \wedge \underset{g}{\star}(d\mathfrak{g}^\beta \wedge \mathfrak{g}_\beta), \tag{7.1}$$

[1] A rigorous formulation of the Lagrangian and Hamiltonian formalism in field theory, needs at least, the introduction of the concepts of jet bundles and the Legendre bundles. Such a theory is well presented, for example., in the excellent texts [1,2]. Here, we give a formulation of the theory, without mentioning those concepts, but which is very similar to the standard one used in physics books on field theory, and which, to the best of our knowledge was used first in [3] and developed with mastery in [4] (see also [5]). Also, we recall that the Hamiltonian formalism for *GRT*, as usually presented on some textbooks, has been introduced in [6], where the concept of *ADM* energy first appeared. See Section 7.3.

V.V. Fernández and W.A. Rodrigues Jr., *Gravitation as a Plastic Distortion of the Lorentz Vacuum*,
Fundamental Theories of Physics 168, DOI 10.1007/978-3-642-13589-7_7,

and as usual in the Lagrangian formalism, we define the conjugate momenta $\mathfrak{p}_\alpha \in \sec \bigwedge^3 T^*M \hookrightarrow \mathcal{C}\ell(\mathbf{g}, M)$ to the potentials $\mathfrak{g}^\alpha$ by:

$$\mathfrak{p}_\alpha = \frac{\partial \mathcal{L}_g}{\partial d\mathfrak{g}^\alpha}, \tag{7.2}$$

and before proceeding, we recall that according to Eq.(6.27), it is:

$$\mathfrak{p}_\alpha \equiv \underset{g}{\star} \mathcal{S}_\alpha. \tag{7.3}$$

Next, supposing that we can solve Eq.(7.2) for the $d\mathfrak{g}^\alpha$ as functions of the $\mathfrak{p}_\alpha$, we introduce a *Legendre* transformation with respect to the fields $d\mathfrak{g}^\alpha$ by

$$\mathbf{L} : (\mathfrak{g}^\alpha, \mathfrak{p}_\alpha) \mapsto \mathbf{L}(\mathfrak{g}^\alpha, \mathfrak{p}_\alpha) = d\mathfrak{g}^\alpha \wedge \mathfrak{p}_\alpha - \mathcal{L}_g(\mathfrak{g}^\alpha, d\mathfrak{g}^\alpha(\mathfrak{p}_\alpha)) \tag{7.4}$$

We write in what follows

$$\mathfrak{L}_g(\mathfrak{g}^\alpha, \mathfrak{p}_\alpha) := \mathcal{L}_g(\mathfrak{g}^\alpha, d\mathfrak{g}^\alpha(\mathfrak{p}_\alpha)) \tag{7.5}$$

We observe that

$$\begin{aligned}
\boldsymbol{\delta}\mathcal{L}_g(\mathfrak{g}^\alpha, d\mathfrak{g}^\alpha) &= d(\boldsymbol{\delta}\mathfrak{g}^\alpha \wedge \frac{\partial \mathcal{L}_g(\mathfrak{g}^\alpha, d\mathfrak{g}^\alpha)}{\partial d\mathfrak{g}^\alpha}) \\
&\quad + \boldsymbol{\delta}\mathfrak{g}^\alpha \wedge \left[ \frac{\partial \mathcal{L}_g(\mathfrak{g}^\alpha, d\mathfrak{g}^\alpha)}{\partial \mathfrak{g}^\alpha} - d(\frac{\partial \mathcal{L}_g(\mathfrak{g}^\alpha, d\mathfrak{g}^\alpha)}{\partial d\mathfrak{g}^\alpha}) \right] \\
&= d(\boldsymbol{\delta}\mathfrak{g}^\alpha \wedge \mathfrak{p}_\alpha) + \boldsymbol{\delta}\mathfrak{g}^\alpha \wedge \frac{\boldsymbol{\delta}\mathcal{L}_g(\mathfrak{g}^\alpha, d\mathfrak{g}^\alpha)}{\boldsymbol{\delta}\mathfrak{g}^\alpha}.
\end{aligned} \tag{7.6}$$

Also from Eq.(7.4) we immediately have:

$$\begin{aligned}
\boldsymbol{\delta}\mathfrak{L}_g(\mathfrak{g}^\alpha, \mathfrak{p}_\alpha) &= \boldsymbol{\delta}(d\mathfrak{g}^\alpha \wedge \mathfrak{p}_\alpha) - \boldsymbol{\delta}\mathfrak{g}^\alpha \wedge \frac{\partial \mathbf{L}}{\partial \mathfrak{g}^\alpha} - \boldsymbol{\delta}\mathfrak{p}_\alpha \wedge \frac{\partial \mathbf{L}}{\partial \mathfrak{p}_\alpha} \\
&= d(\boldsymbol{\delta}\mathfrak{g}^\alpha \wedge \mathfrak{p}_\alpha) + \boldsymbol{\delta}\mathfrak{g}^\alpha \left( -d\mathfrak{p}_\alpha - \frac{\partial \mathbf{L}}{\partial \mathfrak{g}^\alpha} \right) + \boldsymbol{\delta}(d\mathfrak{g}^\alpha - \frac{\partial \mathbf{L}}{\partial \mathfrak{p}_\alpha}) \wedge \mathfrak{p}_\alpha \\
&= d(\boldsymbol{\delta}\mathfrak{g}^\alpha \wedge \mathfrak{p}_\alpha) + \boldsymbol{\delta}\mathfrak{g}^\alpha \wedge \left( \frac{\boldsymbol{\delta}\mathfrak{L}_g(\mathfrak{g}^\alpha, \mathfrak{p}_\alpha)}{\boldsymbol{\delta}\mathfrak{g}^\alpha} \right) \\
&\quad + \left( \frac{\boldsymbol{\delta}\mathfrak{L}_g(\mathfrak{g}^\alpha, \mathfrak{p}_\alpha)}{\boldsymbol{\delta}\mathfrak{p}_\alpha} \right) \wedge \boldsymbol{\delta}\mathfrak{p}_\alpha,
\end{aligned} \tag{7.7}$$

where we put:

$$\begin{aligned}
\frac{\boldsymbol{\delta}\mathfrak{L}_g(\mathfrak{g}^\alpha, \mathfrak{p}_\alpha)}{\boldsymbol{\delta}\mathfrak{g}^\alpha} &:= -d\mathfrak{p}_\alpha - \frac{\partial \mathbf{L}}{\partial \mathfrak{g}^\alpha}, \\
\frac{\boldsymbol{\delta}\mathfrak{L}_g(\mathfrak{g}^\alpha, \mathfrak{p}_\alpha)}{\boldsymbol{\delta}\mathfrak{p}_\alpha} &:= d\mathfrak{g}^\alpha - \frac{\partial \mathbf{L}}{\partial \mathfrak{p}_\alpha}.
\end{aligned} \tag{7.8}$$

From Eqs. (7.6) and (7.7) we have

$$\boldsymbol{\delta}\mathfrak{g}^\alpha \wedge \frac{\boldsymbol{\delta}\mathcal{L}_g(\mathfrak{g}^\alpha, d\mathfrak{g}^\alpha)}{\boldsymbol{\delta}\mathfrak{g}^\alpha} = \boldsymbol{\delta}\mathfrak{g}^\alpha \wedge \left( \frac{\boldsymbol{\delta}\mathfrak{L}_g(\mathfrak{g}^\alpha, \mathfrak{p}_\alpha)}{\boldsymbol{\delta}\mathfrak{g}^\alpha} \right) + \left( \frac{\boldsymbol{\delta}\mathfrak{L}_g(\mathfrak{g}^\alpha, \mathfrak{p}_\alpha)}{\boldsymbol{\delta}\mathfrak{p}_\alpha} \right) \wedge \boldsymbol{\delta}\mathfrak{p}_\alpha. \tag{7.9}$$

To define the Hamiltonian form, we need something to act the role of time for our manifold, and we choose this 'time' to be given by the flow of an arbitrary timelike vector field[2] $\mathbf{Z} \in \sec TM$ such that $\mathbf{g}(\mathbf{Z}, \mathbf{Z}) = 1$. Moreover, we define $Z = \mathbf{g}(\mathbf{Z}, ) \in \sec \bigwedge^1 T^*M \hookrightarrow \mathcal{C}\ell(\mathrm{g}, M)$. With this choice, the variation $\boldsymbol{\delta}$ is generated by the Lie derivative $\pounds_{\mathbf{Z}}$. Using Cartan's 'magical formula' (see, e.g., [7]), we have

$$\boldsymbol{\delta}\mathfrak{L}_g = \pounds_{\mathbf{Z}}\mathfrak{L}_g = d(Z \lrcorner \mathfrak{L}_g) + Z \lrcorner d\mathfrak{L}_g = d(Z \lrcorner \mathfrak{L}_g). \tag{7.10}$$

Then, using Eq.(7.7) we can write

$$d(Z \lrcorner \mathfrak{L}_g) = d(\pounds_{\mathbf{Z}}\mathfrak{g}^\alpha \wedge \mathfrak{p}_\alpha) + \pounds_{\mathbf{Z}}\mathfrak{g}^\alpha \wedge \frac{\boldsymbol{\delta}\mathfrak{L}_g}{\boldsymbol{\delta}\mathfrak{g}^\alpha} + \pounds_{\mathbf{Z}}\mathfrak{p}_\alpha \wedge \left(\frac{\boldsymbol{\delta}\mathfrak{L}}{\boldsymbol{\delta}\mathfrak{p}_\alpha}\right) \tag{7.11}$$

and taking into account Eq.(7.9), we get

$$d(\pounds_{\mathbf{Z}}\mathfrak{g}^\alpha \wedge \mathfrak{p}_\alpha - Z \lrcorner \mathfrak{L}_g) = \pounds_{\mathbf{Z}}\mathfrak{g}^\alpha \wedge \frac{\boldsymbol{\delta}\mathcal{L}_g}{\boldsymbol{\delta}\mathfrak{g}^\alpha}. \tag{7.12}$$

Now, we define the *Hamiltonian* 3-form by

$$\mathcal{H}(\mathfrak{g}^\alpha, \mathfrak{p}_\alpha) := \pounds_{\mathbf{Z}}\mathfrak{g}^\alpha \wedge \mathfrak{p}_\alpha - Z \lrcorner \mathfrak{L}_g. \tag{7.13}$$

We immediately have taking into account Eq.(7.12) that, when the field equations for the *free* gravitational field are satisfied ( i.e., when the Euler-Lagrange functional is null, $\frac{\boldsymbol{\delta}\mathcal{L}_g}{\boldsymbol{\delta}\mathfrak{g}^\alpha} = 0$) that

$$d\mathcal{H} = 0. \tag{7.14}$$

Thus $\mathcal{H}$ is a *conserved* Noether current. We next write

$$\mathcal{H} = Z^\alpha \mathcal{H}_\alpha + dB \tag{7.15}$$

and find some equivalent expressions for $\mathcal{H}_\alpha$. To start, we rewrite Eq.(7.13) (taking into account Eq.(7.2), Cartan's magical formula and some Clifford algebra identities [12]) as

$$\begin{aligned}
\mathcal{H} &= -Z \lrcorner \mathfrak{L}_g + \pounds_{\mathbf{Z}}\mathfrak{g}^\alpha \wedge \mathfrak{p}_\alpha \\
&= -Z \lrcorner \mathfrak{L}_g + d(Z \lrcorner \mathfrak{g}^\alpha) \wedge \mathfrak{p}_\alpha + (Z \lrcorner d\mathfrak{g}^\alpha) \wedge \mathfrak{p}_\alpha \\
&- Z \lrcorner \mathfrak{L}_g + d\left((Z \lrcorner \mathfrak{g}^\alpha) \wedge \mathfrak{p}_\alpha\right) - (Z \lrcorner \mathfrak{g}^\alpha) \wedge d\mathfrak{p}_\alpha + (Z \lrcorner d\mathfrak{g}^\alpha) \wedge \mathfrak{p}_\alpha \\
&= Z^\alpha \left(\mathfrak{g}_\alpha \lrcorner \mathfrak{L}_g + (\mathfrak{g}_\alpha \lrcorner d\mathfrak{g}^\kappa) \wedge \mathfrak{p}_\kappa - d\mathfrak{p}_\alpha\right) + d\left(Z^\alpha \mathfrak{p}_\alpha\right),
\end{aligned} \tag{7.16}$$

from where we get

$$\boxed{\mathcal{H}_\alpha = \mathfrak{g}_\alpha \lrcorner \mathfrak{L}_g + (\mathfrak{g}_\alpha \lrcorner d\mathfrak{g}^\kappa) \wedge \mathfrak{p}_\kappa - d\mathfrak{p}_\alpha,} \\ \boxed{B = Z^\alpha \mathfrak{p}_\alpha.} \tag{7.17}$$

[2] We recall once again that an arbitrary timelike vector field $\mathbf{Z} \in \sec TM$ such that $\mathbf{g}(\mathbf{Z}, \mathbf{Z}) = 1$ defines a *reference frame* in *GRT*, whose integral lines represent *observers*. For details, consult, for example, [7,8].

Next, we use the definition of the Legendre transform (Eq.(7.4)), and some Clifford algebra identities to write Eq.(7.13) as :

$$\begin{aligned}\mathcal{H} &= -Z\lrcorner\mathfrak{L}_g + d(Z\lrcorner\mathfrak{g}^{\alpha})\wedge\mathfrak{p}_{\alpha} + (Z\lrcorner d\mathfrak{g}^{\alpha})\wedge\mathfrak{p}_{\alpha}\\ &= Z\lrcorner\mathbf{L} - Z\lrcorner(\mathfrak{g}^{\kappa}\wedge\mathfrak{p}_{\kappa}) + d(Z\lrcorner\mathfrak{g}^{\kappa})\wedge\mathfrak{p}_{\kappa} + (Z\lrcorner d\mathfrak{g}^{\kappa})\wedge\mathfrak{p}_{\kappa}\\ &= Z^{\alpha}\left(\mathbf{L}_{\alpha} + (\mathfrak{g}_{\alpha}\lrcorner\mathfrak{p}_{\kappa})\wedge d\mathfrak{g}^{\kappa} - (\mathfrak{g}_{\alpha}\lrcorner\mathfrak{g}^{\kappa})d\mathfrak{p}_{\kappa}\right) + d\left(Z^{\alpha}\mathfrak{p}_{\alpha}\right),\end{aligned} \tag{7.18}$$

where we put $\mathbf{L}_{\alpha} = \mathfrak{g}_{\alpha}\lrcorner\mathbf{L}$. Thus, we also have

$$\boxed{\mathcal{H}_{\alpha} = \mathbf{L}_{\alpha} + (\mathfrak{g}_{\alpha}\lrcorner\mathfrak{p}_{\kappa})\wedge d\mathfrak{g}^{\kappa} - d\mathfrak{p}_{\alpha},}$$
$$\boxed{B = Z^{\alpha}\mathfrak{p}_{\alpha}.} \tag{7.19}$$

Now, recalling Eq.(7.12), we can write

$$d(Z^{\alpha}\mathcal{H}_{\alpha} + dB) + \pounds_{\mathbf{Z}}\mathfrak{g}^{\alpha}\wedge\frac{\delta\mathcal{L}_g}{\delta\mathfrak{g}^{\alpha}} = 0. \tag{7.20}$$

Then,

$$\begin{aligned}&dZ^{\alpha}\wedge\mathcal{H}_{\alpha} + Z^{\alpha}d\mathcal{H}_{\alpha} + d(Z\lrcorner\mathfrak{g}^{\alpha})\wedge\frac{\delta\mathcal{L}_g}{\delta\mathfrak{g}^{\alpha}} + (Z\lrcorner d\mathfrak{g}^{\alpha})\wedge\frac{\delta\mathcal{L}_g}{\delta\mathfrak{g}^{\alpha}}\\ &= dZ^{\alpha}\wedge\left(\mathcal{H}_{\alpha} + \frac{\delta\mathcal{L}_g}{\delta\mathfrak{g}^{\alpha}}\right)\\ &+ Z^{\alpha}\left(d\mathcal{H}_{\alpha} + d(\mathfrak{g}_{\alpha}\lrcorner\mathfrak{g}^{\kappa})\wedge\frac{\delta\mathcal{L}_g}{\delta\mathfrak{g}^{\kappa}} + (\mathfrak{g}_{\alpha}\lrcorner d\mathfrak{g}^{\kappa})\wedge\frac{\delta\mathcal{L}_g}{\delta\mathfrak{g}^{\kappa}}\right)\\ &= 0,\end{aligned} \tag{7.21}$$

which means that

$$\mathcal{H}_{\alpha} = -\frac{\delta\mathcal{L}_g}{\delta\mathfrak{g}^{\alpha}} \tag{7.22}$$

and

$$d\mathcal{H}_{\alpha} + d(\mathfrak{g}_{\alpha}\lrcorner\mathfrak{g}^{\kappa})\wedge\frac{\delta\mathcal{L}_g}{\delta\mathfrak{g}^{\kappa}} + (\mathfrak{g}_{\alpha}\lrcorner d\mathfrak{g}^{\kappa})\wedge\frac{\delta\mathcal{L}_g}{\delta\mathfrak{g}^{\kappa}} = 0. \tag{7.23}$$

From Eq.(7.23), we have

$$\begin{aligned}d\mathcal{H}_{\alpha} &= -d(\mathfrak{g}_{\alpha}\lrcorner\mathfrak{g}^{\kappa})\wedge\frac{\delta\mathcal{L}_g}{\delta\mathfrak{g}^{\kappa}} - (\mathfrak{g}_{\alpha}\lrcorner d\mathfrak{g}^{\kappa})\wedge\frac{\delta\mathcal{L}_g}{\delta\mathfrak{g}^{\kappa}}\\ &= -(\mathfrak{g}_{\alpha}\lrcorner d\mathfrak{g}^{\kappa})\wedge\frac{\delta\mathcal{L}_g}{\delta\mathfrak{g}^{\kappa}} - d\left[(\mathfrak{g}_{\alpha}\lrcorner\mathfrak{g}^{\kappa})\wedge\frac{\delta\mathcal{L}_g}{\delta\mathfrak{g}^{\kappa}}\right] + (\mathfrak{g}_{\alpha}\lrcorner\mathfrak{g}^{\kappa})\wedge d\left(\frac{\delta\mathcal{L}_g}{\delta\mathfrak{g}^{\kappa}}\right)\\ &= -(\mathfrak{g}_{\alpha}\lrcorner d\mathfrak{g}^{\kappa})\wedge\frac{\delta\mathcal{L}_g}{\delta\mathfrak{g}^{\kappa}} - d\left[\frac{\delta\mathcal{L}_g}{\delta\mathfrak{g}^{\alpha}}\right] + d\left(\frac{\delta\mathcal{L}_g}{\delta\mathfrak{g}^{\alpha}}\right)\\ &= -(\mathfrak{g}_{\alpha}\lrcorner d\mathfrak{g}^{\kappa})\wedge\frac{\delta\mathcal{L}_g}{\delta\mathfrak{g}^{\kappa}} + d\mathcal{H}_{\alpha} + d\left(\frac{\delta\mathcal{L}_g}{\delta\mathfrak{g}^{\alpha}}\right),\end{aligned} \tag{7.24}$$

from where follows the identity:

$$(\mathfrak{g}_\alpha \lrcorner d\mathfrak{g}^\kappa) \wedge \frac{\delta \mathcal{L}_g}{\delta \mathfrak{g}^\kappa} = d\left(\frac{\delta \mathcal{L}_g}{\delta \mathfrak{g}^\alpha}\right). \tag{7.25}$$

We now return to Eq.(7.20) and recalling Eq.(7.9), we can write:

$$d(Z^\alpha \mathcal{H}_\alpha + dB) + \delta \mathfrak{g}^\alpha \wedge \left(\frac{\delta \mathfrak{L}_g}{\delta \mathfrak{g}^\alpha}\right) + \left(\frac{\delta \mathfrak{L}_g}{\delta \mathfrak{p}_\alpha}\right) \wedge \delta \mathfrak{p}_\alpha = 0, \tag{7.26}$$

from where we get, after some algebra,

$$\boxed{\mathcal{H}_\alpha(\mathfrak{g}^\alpha, \mathfrak{p}_\alpha) = \frac{\delta \mathfrak{L}_g}{\delta \mathfrak{g}^\alpha} - (\mathfrak{g}_\alpha \lrcorner \wedge \mathfrak{p}_\kappa) \wedge \frac{\delta \mathfrak{L}_g}{\delta \mathfrak{p}_\kappa},}$$
$$\boxed{B = Z^\alpha \mathfrak{p}_\alpha.} \tag{7.27}$$

## 7.2 The Quasi Local Energy

Now, let us investigate the meaning of the boundary term in Eq.(7.15). Consider an arbitrary spacelike hypersuface $\sigma$. Then, we define

$$\begin{aligned}\mathbf{H} &= \int_\sigma (Z^\alpha \mathcal{H}_\alpha + dB) \\ &= \int_\sigma Z^\alpha \mathcal{H}_\alpha + \int_{\partial\sigma} B.\end{aligned} \tag{7.28}$$

If we take into account Eq.(7.22), we see that the first term in Eq.(7.22) is null when the field equations (for the free gravitational field) are satisfied and we are thus left with

$$\mathbf{E} = \int_{\partial\sigma} B, \tag{7.29}$$

which is called the quasi local energy[3].

Before proceeding, recall that if $\{\mathfrak{e}_\alpha\}$ is the dual basis of $\{\mathfrak{g}^\alpha\}$ we have

$$\mathfrak{g}^{\mathbf{0}}(\mathfrak{e}_{\mathbf{i}}) = 0, \ \mathbf{i} = 1, 2, 3. \tag{7.30}$$

Then, if we take $\mathbf{Z} = \mathfrak{e}_{\mathbf{0}}$ orthogonal to the hypersurface $\sigma$, such that for each $p \in \sigma$, $T\sigma_p$ is generated by $\{\mathfrak{e}_{\mathbf{i}}\}$, we get recalling that in our theory it is $\mathfrak{p}_\alpha = \underset{g}{\star} \mathcal{S}_\alpha$ (Eq.(6.27)) that

$$\mathbf{E} = \int_{\partial\sigma} \underset{g}{\star} \mathcal{S}_0, \tag{7.31}$$

which we recognize as being the same conserved quantity as the one defined by $P_0'$ in Eq.(6.33), which is the total energy contained in the gravitational plus matter fields for a solution of Einstein's equation, such that the source term goes to zero at spatial infinity.

[3]For an up to date review on the concept of quasilocal energy in *GRT*, its success and drawbacks see [9].

## 7.3 Hamilton's Equations

Before we explore in detail the meaning of this quasi local energy, let us remember that the role of a Hamiltonian density is to derive Hamilton's equation. We have

$$\boldsymbol{\delta}\,\mathbf{H} = \int_\sigma \boldsymbol{\delta}\mathcal{H} = \int_\sigma \boldsymbol{\delta}(Z^\alpha \mathcal{H}_\alpha + dB) \tag{7.32}$$

where $\boldsymbol{\delta}$ is an *arbitrary* variation, not generated by the flow of the vector field $\mathbf{Z}$. To perform that variation, we recall that we obtained above three different (but equivalent) expressions for $\mathcal{H}_\alpha$, namely Eq.(7.17), Eq.(7.19) and Eq.(7.29).

Taking into account that $\boldsymbol{\delta}\mathcal{L}_g(\mathfrak{g}^\alpha, d\mathfrak{g}^\alpha) = \boldsymbol{\delta}\mathfrak{L}_g(\mathfrak{g}^\alpha, \mathfrak{p}_\alpha)$, we have using Eq.(7.6) and Cartan's magical formula[4]

$$\begin{aligned}
\boldsymbol{\delta}\mathcal{H} &= \boldsymbol{\delta}\left(\pounds_{\mathbf{Z}}\mathfrak{g}^\alpha \wedge \mathfrak{p}_\alpha - Z\lrcorner\mathfrak{L}_g\right) \\
&= -Z\lrcorner\left(d(\boldsymbol{\delta}\mathfrak{g}^\alpha \wedge \mathfrak{p}_\alpha) + \boldsymbol{\delta}\mathfrak{g}^\alpha \wedge \frac{\boldsymbol{\delta}\mathcal{L}_g}{\boldsymbol{\delta}\mathfrak{g}^\alpha}\right) + \pounds_{\mathbf{Z}}\boldsymbol{\delta}\mathfrak{g}^\alpha \wedge \mathfrak{p}_\alpha + \pounds_{\mathbf{Z}}\mathfrak{g}^\alpha \wedge \boldsymbol{\delta}\mathfrak{p}_\alpha \\
&= -\boldsymbol{\delta}\mathfrak{g}^\alpha \wedge \pounds_{\mathbf{Z}}\mathfrak{p}_\alpha + \boldsymbol{\delta}\mathfrak{p}_\alpha \wedge \pounds_{\mathbf{Z}}\mathfrak{g}^\alpha - Z\lrcorner(\boldsymbol{\delta}\mathfrak{g}^\alpha \wedge \frac{\boldsymbol{\delta}\mathcal{L}_g}{\boldsymbol{\delta}\mathfrak{g}^\alpha}) + d[Z\lrcorner(\boldsymbol{\delta}\mathfrak{g}^\alpha \wedge \mathfrak{p}_\alpha)].
\end{aligned} \tag{7.33}$$

This last equation can also be written using Eq.(7.9) as

$$\begin{aligned}
\boldsymbol{\delta}\mathcal{H} =& -\boldsymbol{\delta}\mathfrak{g}^\alpha \wedge \pounds_{\mathbf{Z}}\mathfrak{p}_\alpha + \boldsymbol{\delta}\mathfrak{p}_\alpha \wedge \pounds_{\mathbf{Z}}\mathfrak{g}^\alpha \\
&- Z\lrcorner(\boldsymbol{\delta}\mathfrak{g}^\alpha \wedge \frac{\boldsymbol{\delta}\mathfrak{L}_g}{\boldsymbol{\delta}\mathfrak{g}^\alpha} + \boldsymbol{\delta}\mathfrak{p}_\alpha \wedge \frac{\boldsymbol{\delta}\mathfrak{L}_g}{\boldsymbol{\delta}\mathfrak{p}_\alpha}) + d[Z\lrcorner(\boldsymbol{\delta}\mathfrak{g}^\alpha \wedge \mathfrak{p}_\alpha)],
\end{aligned} \tag{7.34}$$

and a simple calculation shows that this is also the form for $\boldsymbol{\delta}\mathcal{H}$ that we get varying Eq.(7.23). From the above results, it follows that

$$\begin{aligned}
\boldsymbol{\delta}\mathbf{H} &= \int_\sigma (-\boldsymbol{\delta}\mathfrak{g}^\alpha \wedge \pounds_{\mathbf{Z}}\mathfrak{p}_\alpha + \boldsymbol{\delta}\mathfrak{p}_\alpha \wedge \pounds_{\mathbf{Z}}\mathfrak{g}^\alpha) \\
&\int_\sigma - Z\lrcorner(\boldsymbol{\delta}\mathfrak{g}^\alpha \wedge \frac{\boldsymbol{\delta}\mathfrak{L}_g}{\boldsymbol{\delta}\mathfrak{g}^\alpha} + \boldsymbol{\delta}\mathfrak{p}_\alpha \wedge \frac{\boldsymbol{\delta}\mathfrak{L}_g}{\boldsymbol{\delta}\mathfrak{p}_\alpha}) \\
&+ \int_{\partial\sigma} Z\lrcorner(\boldsymbol{\delta}\mathfrak{g}^\alpha \wedge \mathfrak{p}_\alpha).
\end{aligned} \tag{7.35}$$

Thus, only if the field equations are satisfied (i.e., $\frac{\boldsymbol{\delta}\mathcal{L}_g}{\boldsymbol{\delta}\mathfrak{g}^\alpha} = 0$ or $\frac{\boldsymbol{\delta}\mathfrak{L}_g}{\boldsymbol{\delta}\mathfrak{g}^\alpha} = 0$, $\frac{\boldsymbol{\delta}\mathfrak{L}_g}{\boldsymbol{\delta}\mathfrak{p}_\alpha} = 0$) and

$$\int_{\partial\sigma} Z\lrcorner(\boldsymbol{\delta}\mathfrak{g}^\alpha \wedge \mathfrak{p}_\alpha) = 0, \tag{7.36}$$

for arbitrary variations $\boldsymbol{\delta}\mathfrak{g}^\alpha$ not necessarily vanishing at the boundary [10] $\partial\sigma$, it follows that the variation $\boldsymbol{\delta}\mathbf{H}$ gives Hamilton's equations:

$$\pounds_{\mathbf{Z}}\mathfrak{p}_\alpha = -\frac{\boldsymbol{\delta}\mathcal{H}}{\boldsymbol{\delta}\mathfrak{g}^\alpha}, \; \pounds_{\mathbf{Z}}\mathfrak{g}^\alpha = \frac{\boldsymbol{\delta}\mathcal{H}}{\boldsymbol{\delta}\mathfrak{p}_\alpha}. \tag{7.37}$$

[4]Keep in mind that the 1-form field physically equivalent to the vector field $\mathbf{Z}$ is $Z$ such that $Z(\mathbf{Z}) = 1$.

Now, the term $d[Z\lrcorner(\boldsymbol{\delta}\mathfrak{g}^{\alpha} \wedge \mathfrak{p}_{\alpha})]$ comes as part of the variation of the term $dB$. Indeed, since

$$\begin{aligned} B &= Z^{\alpha}\mathfrak{p}_{\alpha} = (Z\lrcorner\mathfrak{g}^{\alpha}) \wedge \mathfrak{p}_{\alpha} \\ &= Z\lrcorner(\mathfrak{g}^{\alpha} \wedge \mathfrak{p}_{\alpha}) + \mathfrak{g}^{\alpha} \wedge (Z\lrcorner\mathfrak{p}_{\alpha}), \end{aligned} \tag{7.38}$$

it is

$$dB = d(Z\lrcorner(\mathfrak{g}^{\alpha} \wedge \mathfrak{p}_{\alpha})) + d(\mathfrak{g}^{\alpha} \wedge (Z\lrcorner\mathfrak{p}_{\alpha})), \tag{7.39}$$

and

$$\begin{aligned} \boldsymbol{\delta}dB &= d(Z\lrcorner(\boldsymbol{\delta}\mathfrak{g}^{\alpha} \wedge \mathfrak{p}_{\alpha})) + d(Z\lrcorner(\mathfrak{g}^{\alpha} \wedge \boldsymbol{\delta}\mathfrak{p}_{\alpha})) \\ &+ d(\boldsymbol{\delta}\mathfrak{g}^{\alpha} \wedge (Z\lrcorner\mathfrak{p}_{\alpha})) + d(\mathfrak{g}^{\alpha} \wedge (Z\lrcorner\boldsymbol{\delta}\mathfrak{p}_{\alpha})) \\ &= d(Z\lrcorner(\boldsymbol{\delta}\mathfrak{g}^{\alpha} \wedge \mathfrak{p}_{\alpha})) + d((Z\lrcorner(\mathfrak{g}^{\alpha}) \wedge \boldsymbol{\delta}\mathfrak{p}_{\alpha}) \\ &- \mathfrak{g}^{\alpha} \wedge (Z\lrcorner\boldsymbol{\delta}\mathfrak{p}_{\alpha}) + \mathfrak{g}^{\alpha} \wedge (Z\lrcorner\boldsymbol{\delta}\mathfrak{p}_{\alpha})) + d(\boldsymbol{\delta}\mathfrak{g}^{\alpha} \wedge (Z\lrcorner\mathfrak{p}_{\alpha})) \qquad (7.40) \\ &= d(Z\lrcorner(\boldsymbol{\delta}\mathfrak{g}^{\alpha} \wedge \mathfrak{p}_{\alpha})) + d((Z\lrcorner(\mathfrak{g}^{\alpha}) \wedge \boldsymbol{\delta}\mathfrak{p}_{\alpha}) + d(Z\lrcorner(\boldsymbol{\delta}\mathfrak{g}^{\alpha} \wedge \mathfrak{p}_{\alpha})) \qquad (7.41) \end{aligned}$$

This result is important since it shows that if Eq.(7.36) is not satisfied, we must modify in an appropriate way the boundary term $B$ in Eq.(7.15) in order for that condition to hold. This was exactly the idea originally used by Nester and collaborators [11,12,13] in their proposal for the use of the quasi local energy concept. Here, we analyze what happens in our theory when we choose, e.g., $Z = \mathfrak{g}^{\mathbf{0}}$. So, we examine the integrand in Eq.(7.36). We have:

$$Z\lrcorner(\boldsymbol{\delta}\mathfrak{g}^{\alpha} \wedge \mathfrak{p}_{\alpha}) = (Z\lrcorner\boldsymbol{\delta}\mathfrak{g}^{\alpha}) \wedge \mathfrak{p}_{\alpha} - \boldsymbol{\delta}\mathfrak{g}^{\alpha} \wedge (Z\lrcorner\mathfrak{p}_{\alpha}). \tag{7.42}$$

and with $Z = \mathfrak{g}^{\mathbf{0}}$ it is

$$(Z\lrcorner\boldsymbol{\delta}\mathfrak{g}^{\alpha}) \wedge \mathfrak{p}_{\alpha} = \boldsymbol{\delta}[(Z\lrcorner\mathfrak{g}^{\alpha}) \wedge \mathfrak{p}_{\alpha}] - (\boldsymbol{\delta}Z\lrcorner\mathfrak{g}^{\alpha}) \wedge \mathfrak{p}_{\alpha} - (Z\lrcorner\mathfrak{g}^{\alpha}) \wedge \boldsymbol{\delta}\mathfrak{p}_{\alpha} \tag{7.43}$$

So, if $Z = \mathfrak{g}^{\mathbf{0}}$ is maintained *fixed* on $\partial\sigma$ (for we used this hypothesis in deriving Eq.(7.41), we get

$$Z\lrcorner(\boldsymbol{\delta}\mathfrak{g}^{\alpha} \wedge \mathfrak{p}_{\alpha}) = \boldsymbol{\delta}\mathfrak{p}_{0} - \boldsymbol{\delta}\mathfrak{p}_{0} = 0, \tag{7.44}$$

and thus the boundary term in $\boldsymbol{\delta}\mathbf{H}$ is *null*, thus implying the validity of Hamilton's equations of motion.

**Remark 7.1** If we choose $Z \neq \mathfrak{g}^{\mathbf{0}}$, we will have in general that $Z\lrcorner(\boldsymbol{\delta}\mathfrak{g}^{\alpha} \wedge \mathfrak{p}_{\alpha}) \neq 0$ and, in this case a modification of the boundary term will indeed be necessary in order to get Hamilton's equations of motion. Chen and Nester, making good use of ideas first presented (to the best of our knowledge) in [2] concerning a simplectic framework for field theories, introduced the following prescription for modification of the $B(Z) = (Z\lrcorner\mathfrak{g}^{\alpha}) \wedge \mathfrak{p}_{\alpha} = Z^{\alpha} \wedge \mathfrak{p}_{\alpha}$ term: introduce a reference configuration $(\mathring{\mathfrak{g}}^{\alpha}, \mathring{\mathfrak{p}}_{\alpha})$ and adjust $B(Z)$ to one of the two covariant simplectic forms,

$$B_{\mathfrak{g}^{\alpha}}(Z) = (Z\lrcorner\mathfrak{g}^{\alpha}) \wedge \triangle\mathfrak{p}_{\alpha} + \triangle\mathfrak{g}^{\alpha} \wedge (Z\lrcorner\mathring{\mathfrak{p}}_{\alpha}), \tag{7.45}$$

or

$$B_{\mathfrak{p}_{\alpha}}(Z) = (Z\lrcorner\mathring{\mathfrak{g}}^{\alpha}) \wedge \triangle\mathfrak{p}_{\alpha} + \triangle\mathfrak{g}^{\alpha} \wedge (Z\lrcorner\mathfrak{p}_{\alpha}), \tag{7.46}$$

where $\triangle\mathfrak{g}^{\alpha} = \mathfrak{g}^{\alpha} - \mathring{\mathfrak{g}}^{\alpha}$ and $\triangle\mathfrak{p}_{\alpha} = \mathfrak{p}_{\alpha} - \mathring{\mathfrak{p}}_{\alpha}$. Next, introduce the improved Hamiltonian 3-forms:

$$\mathcal{H}_{\mathfrak{g}^{\alpha}}(Z) = Z^{\mu}\mathcal{H}_{\mu} + dB_{\mathfrak{g}^{\alpha}}(Z), \tag{7.47}$$

and

$$\mathcal{H}_{\mathfrak{p}_\alpha}(Z) = Z^\mu \mathcal{H}_\mu + dB_{\mathfrak{p}_\alpha}(Z). \tag{7.48}$$

For these improved Hamiltonians, we immediately get for arbitrary variations (not generated by the flow of $\mathbf{Z}$)

$$\begin{aligned}\delta\mathcal{H}_{\mathfrak{g}^\alpha}(Z) &= -\delta\mathfrak{g}^\alpha \wedge \pounds_{\mathbf{Z}}\mathfrak{p}_\alpha + \delta\mathfrak{p}_\alpha \wedge \pounds_{\mathbf{Z}}\mathfrak{g}^\alpha \\ &- Z\lrcorner(\delta\mathfrak{g}^\alpha \wedge \frac{\delta\mathfrak{L}_g}{\delta\mathfrak{g}^\alpha} + \delta\mathfrak{p}_\alpha \wedge \frac{\delta\mathfrak{L}_g}{\delta\mathfrak{p}_\alpha}) + d[Z\lrcorner(\delta\mathfrak{g}^\alpha \wedge \triangle\mathfrak{p}_\alpha)],\end{aligned} \tag{7.49}$$

and

$$\begin{aligned}\delta\mathcal{H}_{\mathfrak{p}_\alpha}(Z) &= -\delta\mathfrak{g}^\alpha \wedge \pounds_{\mathbf{Z}}\mathfrak{p}_\alpha + \delta\mathfrak{p}_\alpha \wedge \pounds_{\mathbf{Z}}\mathfrak{g}^\alpha \\ &- Z\lrcorner(\delta\mathfrak{g}^\alpha \wedge \frac{\delta\mathfrak{L}_g}{\delta\mathfrak{g}^\alpha} + \delta\mathfrak{p}_\alpha \wedge \frac{\delta\mathfrak{L}_g}{\delta\mathfrak{p}_\alpha}) - d[Z\lrcorner(\triangle\mathfrak{g}^\alpha \wedge \delta\mathfrak{p}_\alpha)].\end{aligned} \tag{7.50}$$

The variations $\delta\mathcal{H}_{\mathfrak{g}^\alpha}(Z)$ and $\delta\mathcal{H}_{\mathfrak{p}_\alpha}(Z)$ include then, besides the field equations, the following boundary terms,

$$dC_{\mathfrak{g}^\alpha} := d[Z\lrcorner(\delta\mathfrak{g}^\alpha \wedge \triangle\mathfrak{p}_\alpha)], \tag{7.51a}$$
$$-dC_{\mathfrak{p}_\alpha} := -d[Z\lrcorner(\triangle\mathfrak{g}^\alpha \wedge \delta\mathfrak{p}_\alpha)] \tag{7.51b}$$

which are the projections on the boundary of covariant simplectic structures $dC_{\mathfrak{g}^\alpha}$ and $-dC_{\mathfrak{p}_\alpha}$, which simply reflect the choice of the *control* mode according the general methodology introduced in [2].

Now, we can take arbitrary variations on the boundary such that $\delta\mathfrak{g}^\alpha = o(1/r)$ and $\delta\mathfrak{p}_\alpha = o(1/r^2)$ which give $\int_{\partial\sigma} B_{\mathfrak{g}^\alpha}(Z) = 0$ and $\int_{\partial\sigma} B_{\mathfrak{p}_\alpha}(Z) = 0$, warranting the validity of Hamilton's equations. Moreover, we have

$$B_{\mathfrak{g}^\alpha}(Z) - B_{\mathfrak{p}_\alpha}(Z) = Z\lrcorner(\triangle\mathfrak{g}^\alpha \wedge \triangle\mathfrak{p}_\alpha), \tag{7.52}$$

and $\triangle\mathfrak{g}^\alpha \wedge \triangle\mathfrak{p}_\alpha$ is null at spatial infinity if $\mathring{\mathfrak{g}}^\alpha$ and $\mathring{\mathfrak{p}}_\alpha$ are the references concerning the Lorentz vacuum (Minkowski spacetime). Thus, we can use $B_{\mathfrak{g}^\alpha}(Z)$ or $B_{\mathfrak{p}_\alpha}(Z)$ as good boundary terms, but each one gives a different concept of energy (as it happens in Thermodynamics), which may be eventually useful in particular applications.

## 7.4 The *ADM* Energy.

The relation of the previous sections with the original *ADM* formalism [6] can be seen as follows. Instead of choosing an orthonormal vector field $\mathbf{Z}$, start with a global timelike vector field $\mathbf{n} \in \sec TM$ such that $n = \mathbf{g}(\mathbf{N},) = N^2 dt \in \sec \bigwedge^1 T^*M \hookrightarrow \mathcal{C}\ell(M,\mathbf{g})$, with $N : \mathbb{R} \supset \mathbb{I} \rightarrow \mathbb{R}$, a positive function called the lapse function of $M$. Then $n \wedge dn = 0$ and according to Frobenius theorem, $n$ induces a foliation of $M$, i.e., topologically it is $M = \mathbb{I} \times \sigma_t$, where $\sigma_t$ is a spacelike hypersurface with normal given by $\mathbf{n}$. Now, we can decompose any $A \in \sec \bigwedge^p T^*M \hookrightarrow \mathcal{C}\ell(M,\mathbf{g})$ into a tangent component $\underline{A}$ to $\sigma_t$ and an orthogonal component ${}^{\perp}A$ to $\sigma_t$ by [14,15]:

$$A = \underline{A} + {}^{\perp}A, \tag{7.53}$$

where

$$\underline{A} := n\lrcorner(dt \wedge A),\ {}^{\perp}A = dt \wedge A_{\perp} \tag{7.54}$$
$$A_{\perp} := n\lrcorner A. \tag{7.55}$$

Introduce also the parallel component $\underline{d}$ of the differential operator $d$ by:

$$\underline{d}A := dt \wedge (n\lrcorner dA) \tag{7.56}$$

from where it follows (taking into account Cartan's magical formula) that

$$dA = dt \wedge (\pounds_{\mathbf{n}}\underline{A} - \underline{d}A_{\perp}) + \underline{dA}. \tag{7.57}$$

Call

$$\begin{aligned} \boldsymbol{m} &:= -\boldsymbol{g} + n \otimes n \\ &= \underline{\mathfrak{g}}^{i} \otimes \underline{\mathfrak{g}}_{i}, \end{aligned} \tag{7.58}$$

the first fundamental form on $\sigma_t$ and next introduce the Hodge dual operator associated[5] to $\boldsymbol{m}$, acting on the (horizontal forms) forms $\underline{A}$ by

$$\underset{m}{\star}\underline{A} := \underset{g}{\star}(\frac{n}{N} \wedge \underline{A}). \tag{7.59}$$

At this point, we come back to the Lagrangian density Eq.(7.13) and, proceeding like in Sections 7.1 and 7.2, but now leaving $\boldsymbol{\delta} n^{\alpha}$ to be non null, we arrive at the following Hamiltonian density

$$\mathcal{H}(\underline{\mathfrak{g}}^{i}, \underline{\mathfrak{p}_i}) = \pounds_{\mathbf{n}}\underline{\mathfrak{g}}^{i} \wedge \underset{m}{\star}\underline{\mathfrak{p}_i} - \mathcal{K}_g, \tag{7.60}$$

where

$$\overset{\perp}{\mathfrak{g}}{}^{i} = dt \wedge (n\lrcorner \mathfrak{g}^{i}) = n^{i}dt, \tag{7.61}$$

and where $K_g$ depends on $(n, \underline{d}n, \underline{\mathfrak{g}}^{i}, \underline{d\mathfrak{g}}^{i}, \pounds_{\mathbf{n}}\underline{\mathfrak{g}}^{i})$. As in Section 7.1, we can show (after some tedious but straightforward algebra[6]) that $\mathcal{H}(\underline{\mathfrak{g}}^{i}, \underline{\mathfrak{p}_i})$ can be put into the form

$$\mathcal{H} = n^{i}\mathcal{H}_i + \underline{d}B', \tag{7.62}$$

with

$$\mathcal{H}_i = \frac{\boldsymbol{\delta}\mathcal{L}_g}{\boldsymbol{\delta}\mathfrak{g}^{i}} = \frac{\boldsymbol{\delta}\mathcal{K}_g}{\boldsymbol{\delta} n^{i}} \tag{7.63}$$

and

$$B' = -N\underline{\mathfrak{g}_i} \wedge \underset{m}{\star}\underline{d\mathfrak{g}^{i}} \tag{7.64}$$

Then, on shell, i.e., when the field equations are satisfied we get

$$\mathbf{E}' = -\int_{\partial\sigma_t} N\underline{\mathfrak{g}_i} \wedge \underset{m}{\star}\underline{d\mathfrak{g}^{i}}, \tag{7.65}$$

[5]Recall that $\sigma$ has an internal orientation compatible with the orientation of $M$.
[6]Details may be found in [4].

which is exactly the *ADM* energy, as can be seen if we take into account that taking $\partial\sigma_t$ as a twosphere at infinity, we have (using coordinates in the ELP gauge) $\underline{\mathfrak{g}_i} = h_{ij}\underline{d}\mathrm{x}^j$ and $h_{ij}$, $N \to 1$. Then

$$\underline{\mathfrak{g}_i} \wedge \underset{m}{\underline{\star d\mathfrak{g}^i}} = h^{ij}(\frac{\partial h_{ij}}{\partial \mathrm{x}^k} - \frac{\partial h_{ik}}{\partial \mathrm{x}^j}) \underset{m}{\star} \underline{\mathfrak{g}^k} \tag{7.66}$$

and under the above conditions we have the *ADM* formula

$$\mathbf{E}' = \int_{\partial\sigma_t} \left( \frac{\partial h_{ik}}{\partial \mathrm{x}^i} - \frac{\partial h_{ii}}{\partial \mathrm{x}^k} \right) \underset{m}{\star} \underline{\mathfrak{g}^k}, \tag{7.67}$$

which, as is well known[7], is positive definite [15]. When we choose $n = \mathfrak{g}^0$ as in the previous section, we see that $\mathbf{E}' = \mathbf{E}$.

# References

[1] Giachetta, G., Mangiarotti L., and Sardanashvily G., *New Lagrangian and Hamiltonian Methods in Field Theory*, World Sci. Publ. Co., Singapore, 1997.

[2] Kiwjowski, J., and Tulczyjew, M., *A Simpletic Framework for Field Theories*, Lecture Notes in Physics **107**, Springer-Verlag, Berlin, 1970.

[3] Thirring, W., and Wallner, R., The Use of Exterior Forms in Einstein's Gravitational Theory, *Brazilian J. Phys.* **8**, 686-723 (1978).

[4] Wallner, R. P., Asthekar's Variables Reexamined, *Phys. Rev. D.* **46**, 4263-4285 (1992).

[5] Kopczynski, W., Variational Principles for Gravity and Fluids, *Ann. Phys.* **203**, 308-338 (1990).

[6] Arnowitt, B., Deser, S., and Misner, C. W., The Dynamics of General Relativity, in Witten, L. (ed.), *Gravitation, an Introduction to Current Research*, pp. 227-265, J. Willey &Sons, New York, 1962. [http://arXiv.org/abs/gr-qc/0405109v1]

[7] Rodrigues, W. A. Jr., and Capelas de Oliveira, E., T*he Many Faces of Maxwell, Dirac and Einstein Equations. A Clifford Bundle Approach.* Lecture Notes in Physics **722**, Springer, Heidelberg, 2007.

[7] Sachs, R. K., and Wu, H., *General Relativity for Mathematicians*, Springer-Verlag, New York 1977.

[8] Szabados, L. B., Quasi-Local Energy-Momentum and Angular Momentum in GR: A Review Article, *Living Reviews in Relativity*, [http://www.livingreviews.org/lrr-2004-4]

[9] Regge, T., and Teitelboim, C., Role of Surface Integrals in the Hamiltonan Formulation of General Relativity, *Ann. Phys.* **88**, 286-318 (1974).

[10] Chang, C.-C, Nester, J. M., and Chen, C.-C, Pseudotensors and Quasilocal Energy-Momentum, *Phys. Rev. Lett.* **83**, 1897-1901 (1999).

[11] Chen, C.-M., and Nester, J. M., A Simpletic Hamiltonian Derivation of Quasilocal Energy-Momentum for GR, *Grav. Cosmol.* **6**, 257–270 (2000).

[12] Meng, F.-F., *Quasilocal Center-of-Mass Moment in General Relativity*, MSc. thesis, National Central University, Chungli, 2001. [http://thesis.lib.ncu.edu.tw/ETD-db/ETD-search/view_etd?URN=89222030]

[7] As shown by Wallner [4] the formalism used in this section permits an alternative (and very simple) proof of the positive mass theorem, different from the one in [15].

[13] Hecht, R. D., Lemke, J., and Wallner, R. P., Can Poincaré Gauge Theory be Saved?, *Phys. Rev. D* **44**, 2442-2451 (1991)
[14] Witten, E., A New Proof of the Positive Energy Theorem, *Commun. Math. Phys.* **30**, 381-402 (1981).

# 8

# Conclusions

In this book, we presented a theory of the gravitational field, where that field is mathematically represented by a $(1,1)$-extensor field $\boldsymbol{h}$ existing in Minkowski spacetime $(M, \boldsymbol{\eta}, D, \tau_\eta, \uparrow)$ and describes a plastic distortion of a medium that we called the Lorentz vacuum. We showed (Remark 6.2) that the presence of a nontrivial distortion field can be *described* by distinct effective geometrical structures $(M = \mathbb{R}^4, \boldsymbol{g} = \boldsymbol{h}\eta\boldsymbol{h}^\dagger, \nabla, \tau_g)$, which may be associated to a Lorentzian spacetime, or a teleparallel spacetime or even a more general Riemann-Cartan-Weyl spacetime[1]. We postulated a Lagrangian density for the distortion field in interaction with matter, and developed with details the mathematical theory (the multiform and extensor calculus) necessary to obtain the equations of motion of the theory from a variational principle. Moreover, we showed that, when our theory is written in terms of the global potentials $\mathfrak{g}^\alpha = \boldsymbol{h}(\boldsymbol{\vartheta}^\alpha) \in \sec \overset{1}{\bigwedge} T^*M$, $\alpha = 0, 1, 2, 3$, obtained by the deformation of the (continuum) cosmic lattice $\{\boldsymbol{\vartheta}^\alpha\}$, describing the Lorentz vacuum in its ground state, the Lagrangian density contains a term of the Yang-Mills type plus a gauge fixing term and plus a term describing the interaction of the 'vorticities' of the potentials $\mathfrak{g}^\alpha$, and, in this case, no *connection* associated with any geometrical structure appears in it. We proved that, in our theory (in contrast with *GRT*), there are trustworthy conservation laws of energy-momentum and angular momentum for the system consisting of the gravitational plus matter fields. We showed, moreover, how the results obtained for the energy-momentum conservation law from the Lagrangian formalism appear directly in the Hamiltonian formalism of the theory, and how the energy obtained through the Hamiltonian formalism is also related to the concept of *ADM* energy used in *GRT*. We think that those results are enough to clearly illustrate how our theory and *GRT* differ, but, of

[1]In that latter case, we can get equations of motion similar to the Einstein equations, but where the matter tensor appears as a combination of terms proportional to the torsion tensor, that, as well known, in continuum theories describe some special defects in the medium. This possibility will be discussed with more details in another publication. For a simple model where the gravitational field is represented by the nonmetricity tensor of a connection, see Appendix G.

V.V. Fernández and W.A. Rodrigues Jr., *Gravitation as a Plastic Distortion of the Lorentz Vacuum*,
Fundamental Theories of Physics 168, DOI 10.1007/978-3-642-13589-7_8,

course, there are also other crucial differences. In our theory, the spacetime topology is fixed (it is diffeomeorphic to $\mathbb{R}^4$). This implies that in our theory there are no spacetime "singularities", there are no blackholes[2], wormholes, time machines, etc..., but the fact is that there is no *convincing* evidence for any of those exotics objects. We also discussed in Appendix F why a gauge theory of gravitation described in [3,4], and apparently similar to ours does not work. To end, we would like to observe that eventually some readers will not fell happy seeing us to introduce an universal medium filling all spacetime, because it eventually looks like the revival of the aether of the $XIX^{th}$ century. To these people, we emphasize that our theory does *not* violate Lorentz invariance, since the medium and its distortions are represented by a $(1,1)$-extensor field $\boldsymbol{h}$, which does not fix any vector field (or 1-form field). Indeed, any timelike global potential $\mathfrak{g}^0$, in a set of potentials $\{\mathfrak{g}^\alpha\}$, does not fix any privileged reference frame, since the theory is gauge invariant, i.e., the Lagrangian densitiy of the theory is invariant under local Lorentz transformations of the $\mathfrak{g}^\alpha$. So, the medium (the Lorentz vacuum) in our theory has the same role as the one introduced by Einstein in his not so well known 1922 Leyden lecture [4]. Moreover, there are now many arguments for the existence of such a medium (that quantum physicists call quantum vacuum) in, e.g., [6,7,8,9,10,11,12,13]. Eventually, it is also worth quoting here [14,15] which show that if we interpret particles as defects in the medium[3], then some characteristic relativistic and quantum field theory properties of those objects emerge in a natural way. Moreover, there is possibly some hope of describying the electromagnetic field as some kind of distortion of that medium.We intend to come back to these issues in a forthcoming book. Finally, we will be happy if we have motivated the reader to seriously consider the idea that it is, eventually, time to deconstruct the idea that gravitation must necessarily be associated with the geometry of spacetime, and look for its real nature.

# References

[1] Logunov, A. A., and, Mestvirishvili, M., *The Relativistic Theory of Gravitation*, Mir Publishers, Moscow, 1989.

[2] da Rocha, R., and, Rodrigues, W. A. Jr., *Gauge Fixing in the Maxwell Like Graviational Theory in Minkowski Spacetime and in the Equivalent Lorentzian Spacetime.* [arXiv:0807.3275]

[3] Doran, C., and, Lasenby, A., *Geometric Algebra for Physicists*, Cambridge Univ. Press, Cambridge 2003.

[4] Hestenes, D., Gauge Theory Gravity with Geometrical Calculus, *Found. Phys.* **35**, 903-969 (2005).

[5] Einstein, A., Ether and the Theory of Relativity, an address delivered on May 5th, 1920, in the University of Leyden. [http://www.tu-harburg.de/rzt/rzt/it/Ether.html]

[6] Bjorken, J. D., The New Ether, in Fesbach, H., Matsui, T. and Oleson, A., (eds.), *Niels Borh: Physics and the World*, Harwood Acad. Publ., 1988.

[2]This statement means that gravitational collapse does not produce any spacetime singularity. Besides that, it can be proved that if we solve the field equations of our theory for a situation describying gravitational collapse using a specific gauge condition for the gravitational field as done in [1] (something permited in our theory, which does not fix, a priori, any gauge condition [2]) any gravitational collapse stop as the density of matter arrives at a *finite* maximum.

[3]These topological defects introduce obstructions for the Pfaff topology generated by the $\mathfrak{g}^\alpha$ [16,17], a subject that we shall discuss in another publication.

[7] Dirac, P. A. M., Is there an Aether? *Nature* 168, 906-907 (1951).
[8] Laughlin, R.B., A Different Universe: Reinventing Physics from the Bottom, Basic Books, New York, 2005.
[9] Lee, T. D., *Vacuum as the Source of Asymmetry, Particle Physics and Introduction to Field Theory*, Chap. 16, Harwood Acad. Publ., New York,1981.
[10] Schmelzer, I., *A Generalization of the Lorentz Ether to Gravity with General -Relativistic Limit.* [arXiv:gr-qc/0205035]
[11] Schmelzer, I., A Condensed Matter Interpretation of SM Fermions and Gauge Fields, *Found. Phys.* **39**, 73-107 (2009). [arXiv:0908.0591]
[12] Schmelzer, I., *The Background as a Quantum Observable: Einstein's Hole Argument in a Quasiclassical Context.* [arXiv:0909.1408]
[13] Volovik, G. E., *The Universe in a Helium Droplet*, Clarendon Press, Oxford, 2003.
[14] Unzicker, A., What Can Physics Learn from Continuum Mechanics, [arXiv:gr-qc/0011064v1]
[15] Unzicker, A., Teleparallel Space-Time with Defects yields Geometrization of Electrodynamics with Quantized Charges. [arXiv:gr-qc/9612061v2]
[16] Kiehn R. M, *A topological theory of the Physical Vacuum.* [arXiv:gr-qc/0602118]
[17] Kiehn R. M., *The Cosmological Vacuum from a Topological Perspective Part I.* [arXiv:0712.1180]

# Appendix A

## May a Torus with Null Riemann Curvature Exist on $\mathbb{E}^3$?

We can also give a flat Riemann curvature tensor for[1] $\mathcal{T}_3 \simeq S^1 \times S^1 \simeq \mathcal{T}_1 \simeq \mathcal{T}_2$. First parametrize $\mathcal{T}_3$ by $(x^1, x^2)$, $x^1, x^2 \in \mathbb{R}$, such that its coordinates in $\mathbb{R}^3$ are $\mathbf{x}(x^1, x^2) = \mathbf{x}(x^1 + 2\pi, x^2 + 2\pi)$ and

$$\mathbf{x}(x^1, x^2) = (h(x^1)\cos x^2, h(x^1)\sin x^2, l(x^1)), \tag{A.1}$$

with

$$h(x^1) = R + r\cos x^1,\ l(x^1) = r\sin x^1, \tag{A.2}$$

where *here* $R$ and $r$ are real positive constants and $R > r$ (See Figure A.1).

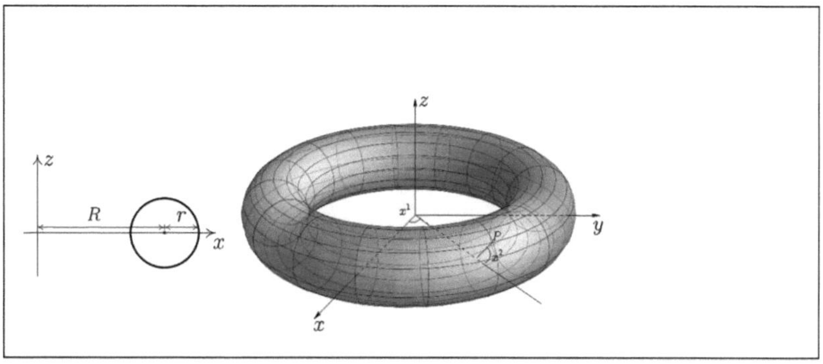


Figure A.1: Some Possible *GSS* for the Torus $\mathcal{T}_3$ living in $\mathbb{E}^3$ where the grid defines the parallelism

[1] Recall that $\mathcal{T}_1$ and $\mathcal{T}_2$ have been defined in Section 1.1.

Now, $T\mathcal{T}_3$ has a global coordinate basis $\{\partial/\partial x^1, \partial/\partial x^2\}$, and the induced metric on $\mathcal{T}_3$ is easily found as

$$\mathbf{g} = r^2 dx^1 \otimes dx^1 + (R + r\cos x^1)^2 dx^2 \otimes dx^2 \tag{A.3}$$

Now, let us introduce a connection $\nabla$ on $\mathcal{T}_3$ such that

$$\nabla_{\partial/\partial x^i} \partial/\partial x^j = 0. \tag{A.4}$$

With respect to this connection, it is immediately verified that its torsion and curvature tensors are null, but the *nonmetricity* of the connection is non null. Indeed,

$$\nabla_{\partial/\partial x^i} \mathbf{g} = -2(R + r\cos x^1)\sin x^1. \tag{A.5}$$

So, in this case, we have a particular Riemann-Cartan-Weyl *GSS* $(\mathcal{T}_3, \mathbf{g}, \nabla)$ with $\mathfrak{A} \neq 0$, $\Theta = 0$, $\mathfrak{R} = 0$.

We can alternatively define the parallelism rule on $\mathcal{T}_3$ by introducing a metric compatible connection $\nabla'$ as follows. First, we introduce the global orthonormal basis $\{\mathbf{e}_i\}$ for $T\mathcal{T}_3$ with

$$\mathbf{e_1} = \frac{1}{r}\frac{\partial}{\partial x^1},\ \mathbf{e_2} = \frac{1}{(R + r\cos x^1)}\frac{\partial}{\partial x^2}, \tag{A.6}$$

and, then, postulate that

$$\nabla'_{\mathbf{e}_i} \mathbf{e_j} = 0. \tag{A.7}$$

Since

$$[\mathbf{e_1}, \mathbf{e_2}] = \frac{r\sin x^1}{r(R + r\cos x^1)}\mathbf{e_2} \tag{A.8}$$

we immediately get that the torsion operator of $\nabla'$ is

$$\overset{\nabla'}{\tau}(\mathbf{e_1}, \mathbf{e_2}) = [\mathbf{e_1}, \mathbf{e_2}] = \frac{R\sin x^1}{r(R + r\cos x^1)}\mathbf{e_2}, \tag{A.9}$$

and the non null components of the torsion tensor are $T^{\mathbf{2}}_{\mathbf{12}}$ and $T^{\mathbf{2}}_{\mathbf{21}} = -T^{\mathbf{2}}_{\mathbf{12}}$,

$$\begin{aligned} T^{\mathbf{2}}_{\mathbf{12}} &= \overset{\nabla'}{\Theta}(\theta^{\mathbf{2}}, \mathbf{e_1}, \mathbf{e_2}) = -\theta^{\mathbf{2}}(\tau(\mathbf{e_1}, \mathbf{e_2})) = -\theta^{\mathbf{2}}\left(-[\mathbf{e_1}, \mathbf{e_2}]\right) \\ &= -\frac{R\sin x^1}{r(R + r\cos x^1)}. \end{aligned} \tag{A.10}$$

Also, it is trivial to see that the curvature operator is null and that $\nabla'\mathbf{g} = 0$, and, in this case, we have a particular Riemann-Cartan *MCGSS* $(\mathcal{T}_3, \mathbf{g}, \nabla')$, with $\Theta \neq 0$ and with a null curvature tensor.

# Appendix B

## Levi-Civita and Nunes Connections on $\mathring{S}^2$

First, consider $S^2$, a sphere of radius $R = 1$ embedded in $\mathbb{R}^3$. Let $(x^1, x^2) = (\vartheta, \varphi)$ $0 < \vartheta < \pi$, $0 < \varphi < 2\pi$, be the standard spherical coordinates of $S^2$, which cover all the open set $U$ which is $S^2$ with the *exclusion* of a semi-circle uniting the *north* and south *poles.*

Introduce the *coordinate bases*

$$\{\boldsymbol{\partial}_\mu\}, \{\theta^\mu = dx^\mu\} \tag{B.1}$$

for $T^*U$ and $T^*U$. Next, introduce the *orthonormal bases* $\{\mathbf{e_a}\}, \{\theta^{\mathbf{a}}\}$ for $T^*U$ and $T^*U$ with

$$\mathbf{e_1} = \boldsymbol{\partial}_1,\ \mathbf{e_2} = \frac{1}{\sin x^1}\boldsymbol{\partial}_2, \tag{B.2a}$$

$$\theta^{\mathbf{1}} = dx^1,\ \theta^{\mathbf{2}} = \sin x^1 dx^2. \tag{B.2b}$$

Then,

$$\begin{aligned}[\mathbf{e_i}, \mathbf{e_j}] &= c_{\mathbf{ij}}^{\mathbf{k}} \mathbf{e_k}, \\ c_{\mathbf{12}}^{\mathbf{2}} &= -c_{\mathbf{21}}^{\mathbf{2}} = -\cot x^{\mathbf{1}}.\end{aligned} \tag{B.3}$$

Moreover, the metric $\boldsymbol{g} \in \sec T_2^0 S^2$ inherited from the ambient Euclidean metric is:

$$\begin{aligned}\boldsymbol{g} &= dx^1 \otimes dx^1 + \sin^2 x^1 dx^2 \otimes dx^2 \\ &= \theta^{\mathbf{1}} \otimes \theta^{\mathbf{1}} + \theta^{\mathbf{2}} \otimes \theta^{\mathbf{2}}.\end{aligned} \tag{B.4}$$

The Levi-Civita connection $D$ of $\mathbf{g}$ has the following non null connections coefficients $\Gamma_{\mu\nu}^{\rho}$ in the coordinate basis (just introduced):

$$\begin{aligned}&D_{\boldsymbol{\partial}_\mu} \boldsymbol{\partial}_\nu = \Gamma_{\mu\nu}^{\rho} \boldsymbol{\partial}_\rho, \\ &\Gamma_{21}^2 = \Gamma_{\theta\varphi}^{\varphi} = \Gamma_{12}^2 = \Gamma_{\varphi\theta}^{\varphi} = \cot\vartheta,\ \Gamma_{22}^1 = \Gamma_{\varphi\varphi}^{\vartheta} = -\cos\vartheta \sin\vartheta.\end{aligned} \tag{B.5}$$

Also, in the basis $\{\mathbf{e_a}\}$, $D_{\mathbf{e_i}}\mathbf{e_j} = \omega_{\mathbf{ij}}^{\mathbf{k}}\mathbf{e_k}$ and the non null coefficients are:

$$\omega_{\mathbf{21}}^{\mathbf{2}} = \cot\vartheta,\ \omega_{\mathbf{22}}^{\mathbf{1}} = -\cot\vartheta. \tag{B.6}$$

The torsion and the (Riemann) curvature tensors of $D$ are

$$\Theta(\theta^{\mathbf{k}}, \mathbf{e_i}, \mathbf{e_j}) = \theta^{\mathbf{k}}(\tau(\mathbf{e_i}, \mathbf{e_j})) = \theta^{\mathbf{k}}\left(D_{\mathbf{e_j}}\mathbf{e_i} - D_{\mathbf{e_i}}\mathbf{e_j} - [\mathbf{e_i}, \mathbf{e_j}]\right), \tag{B.7}$$

$$\mathbf{R}(\mathbf{e_k}, \theta^{\mathbf{a}}, \mathbf{e_i}, \mathbf{e_j}) = \theta^{\mathbf{a}}\left(\left[D_{\mathbf{e_i}}D_{\mathbf{e_j}} - D_{\mathbf{e_j}}D_{\mathbf{e_i}} - D_{[\mathbf{e_i},\, \mathbf{e_j}]}\right]\mathbf{e_k}\right), \tag{B.8}$$

which results in $\mathcal{T} = 0$ and the non null components of $\mathbf{R}$ are $R_{\mathbf{1}\ \mathbf{21}}^{\ \mathbf{1}} = -R_{\mathbf{1}\ \mathbf{12}}^{\ \mathbf{1}} = R_{\mathbf{1}\ \mathbf{12}}^{\ \mathbf{2}} = -R_{\mathbf{1}\ \mathbf{12}}^{\ \mathbf{2}} = -1$.

Since the Riemann curvature tensor is non null, the parallel transport of a given vector depends on the path to be followed. We say that a vector (say $\mathbf{v}_0$) is parallel transported along a generic path $\mathbb{R} \supset I \mapsto \gamma(s) \in \mathbb{R}^3$ (say, from $A = \gamma(0)$ to $B = \gamma(1)$) with tangent vector $\gamma_*(s)$ (at $\gamma(s)$) if it determines a vector field $\mathbf{V}$ along $\gamma$ satisfying

$$D_{\gamma_*}\mathbf{V} = 0, \tag{B.9}$$

and such that $\mathbf{V}(\gamma(0)) = \mathbf{v}_0$. When the path is a geodesic[1] of the connection $D$, i.e., a curve $\mathbb{R} \supset I \mapsto c(s) \in \mathbb{R}^3$ with tangent vector $c_*(s)$ (at $c(s)$) satisfying

$$D_{c_*}c_* = 0, \tag{B.10}$$

the parallel transported vector along $c$ forms a constant angle with $c_*$. Indeed, from Eq.(B.9) it is $\gamma_* \underset{g}{\cdot} D_{\gamma_*}\mathbf{V} = 0$. Then, taking into account Eq.(B.10) it follows that

$$D_{\gamma_*}(\gamma_* \underset{g}{\cdot} \mathbf{V}) = 0.$$

i.e., $\gamma_* \underset{g}{\cdot} \mathbf{V} = \textit{constant}$. This is clearly illustrated in Figure B.1 (adapted from [1]).

Consider next the manifold $\mathring{S}^2 = \{S^2\backslash\text{north pole} + \text{south pole}\} \subset \mathbb{R}^3$, which is a sphere of unitary radius $R = 1$, but this time *excluding* the north and south poles. Let again $\mathbf{g} \in \sec T_2^0\mathring{S}^2$ be the metric field on $\mathring{S}^2$ inherited from the ambient space $\mathbb{R}^3$ and introduce on $\mathring{S}^2$ the Nunes (or navigator) connection $\nabla$ defined by the following parallel transport rule: a vector at an arbitrary point of $\mathring{S}^2$ is parallel transported along a curve $\gamma$, if it determines a vector field on $\gamma$, such that, at any point of $\gamma$ the angle between the transported vector and the vector tangent to the latitude line passing through that point is constant during the transport. This is clearly illustrated in Figure B.2. And to distinguish the Nunes transport from the Levi-Civita transport, we also ask the reader to study the caption of Figure B.1.

We recall that, from the calculation of the Riemann tensor $\mathbf{R}$, it follows that the structures $(\mathring{S}^2, \mathbf{g}, D, \tau_g)$ and also $(S^2, \mathbf{g}, D, \tau_g)$ are Riemann spaces of *constant curvature*. We now show that

[1] We recall that a geodesic of $D$ also determines the minimal distance (as given by the metric $\mathbf{g}$) between any two points on $S^2$.

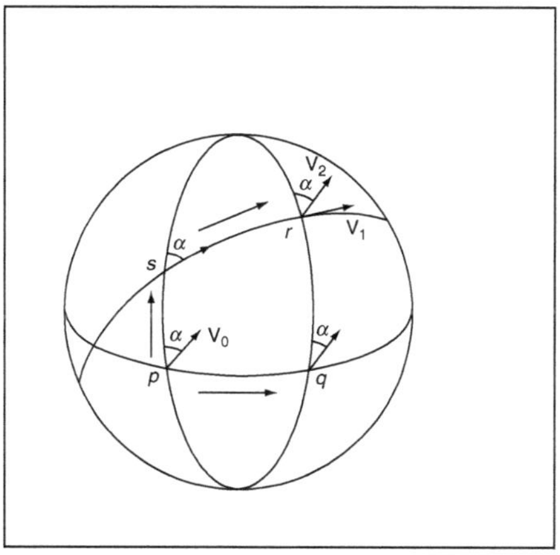


Figure B.1: Levi-Civita and Nunes transport of a vector $\mathbf{v}_0$ starting at $p$ through the paths $psr$ and $pqr$. Levi-Civita transport through psr leads to $\mathbf{v}_1$ whereas Nunes transport leads to $\mathbf{v}_2$. Along $pqr$ both Levi-Civita and Nunes transport agree and leads to $\mathbf{v}_2$.

the structure $(\mathring{S}^2, \mathbf{g}, \nabla, \tau_g)$ is a *teleparallel space*[2], with zero Riemann curvature tensor, but non zero torsion tensor.

Indeed, from Figure B.2, it is clear that (a) if a vector is transported along the *infinitesimal* quadrilateral $pqrs$ composed of latitudes and longitudes, first starting from $p$ along $pqr$, and then starting from $p$ along $psr$, the parallel transported vectors that result in both cases will coincide (study also the caption of Figure B.2).

Now, the vector fields $\mathbf{e_1}$ and $\mathbf{e_2}$ in Eq.(B.2a) define a basis for each point $p$ of $T_p\mathring{S}^2$ and $\nabla$ is clearly characterized by:

$$\nabla_{\mathbf{e_j}}\mathbf{e_i} = 0. \tag{B.11}$$

The components of the curvature operator are:

$$\overset{\nabla}{\mathbf{R}}(\mathbf{e_k}, \theta^{\mathbf{a}}, \mathbf{e_i}, \mathbf{e_j}) = \theta^{\mathbf{a}}\left(\left[\nabla_{\mathbf{e_i}}\nabla_{\mathbf{e_j}} - \nabla_{\mathbf{e_j}}\nabla_{\mathbf{e_i}} - \nabla_{[\mathbf{e_i},\mathbf{e_j}]}\right]\mathbf{e_k}\right) = 0, \tag{B.12}$$

and the torsion operator is

$$\begin{aligned}\overset{\nabla}{\tau}(\mathbf{e_i}, \mathbf{e_j}) &= \nabla_{\mathbf{e_j}}\mathbf{e_i} - \nabla_{\mathbf{e_i}}\mathbf{e_j} - [\mathbf{e_i}, \mathbf{e_j}] \\ &= [\mathbf{e_i}, \mathbf{e_j}],\end{aligned} \tag{B.13}$$

which gives for the components of the torsion tensor, $T^{\mathbf{2}}_{\mathbf{12}} = -T^{\mathbf{2}}_{\mathbf{12}} = \cot\vartheta$. It follows that $\mathring{S}^2$, considered as part of the structure $(\mathring{S}^2, \mathbf{g}, \nabla, \tau_g)$, is *flat* (but has torsion)!

If you need more details to grasp this last result, consider Figure B.2 (b), which shows the standard parametrization of the points $p, q, r, s$ in terms of the spherical coordinates introduced above. According to the geometrical meaning of torsion, its value at a given point is determined

[2]As recalled in Section 1, a teleparallel manifold $M$ is characterized by the existence of global vector fields, which is a basis for $T_xM$ for any $x \in M$. The reason for considering $\mathring{S}^2$ for introducing the Nunes connection is that, as is well known (see, e.g., [2]), $S^2$ does not admit a continuous vector field that is nonnull at all its points.

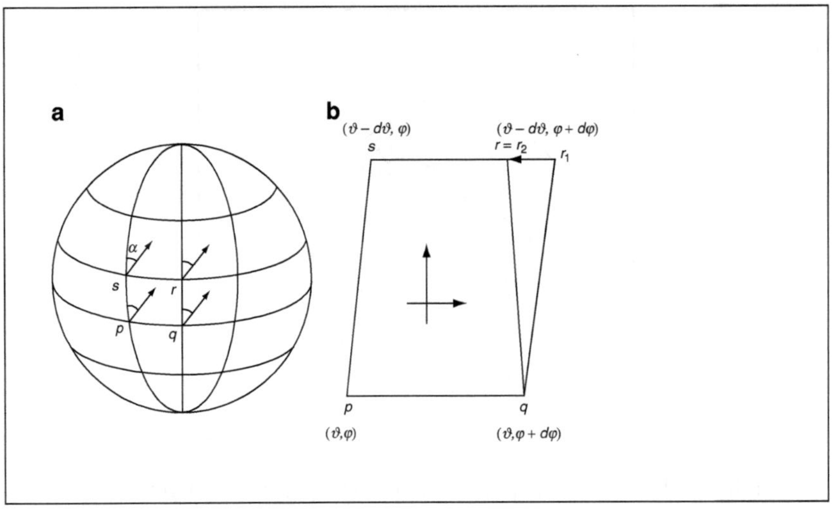


Figure B.2: Characterization of the Nunes connection.

by calculating the difference between the (infinitesimal)[3] vectors $pr_1$and $pr_2$. If the vector $pq$ is transported along $ps$, one gets (recalling that $R = 1$) the vector $\mathbf{v} = sr_1$ such that $|\mathbf{g}(\mathbf{v}, \mathbf{v})|^{\frac{1}{2}} = \sin\vartheta\triangle\varphi$. On the other hand, if the vector $ps$ is transported along $pq$, one gets the vector $qr_2 = qr$. Let $\mathbf{w} = sr$. Then,

$$|\mathbf{g}(\mathbf{w}, \mathbf{w})| = \sin(\vartheta - \triangle\vartheta)\triangle\varphi \simeq \sin\vartheta\triangle\varphi - \cos\vartheta\triangle\vartheta\triangle\varphi, \tag{B.14}$$

Also,

$$\mathbf{u} = r_1r_2 = -u(\frac{1}{\sin\vartheta}\frac{\partial}{\partial\varphi}),\ u = |\mathbf{g}(\mathbf{u}, \mathbf{u})| = \cos\vartheta\triangle\vartheta\triangle\varphi. \tag{B.15}$$

Then, the connection $\nabla$ of the structure $(\mathring{S}^2, \mathbf{g}, \nabla, \tau_g)$ has a non null torsion tensor $\bar{T}$. Indeed, the component of $\mathbf{u} = r_1r_2$ in the direction $\partial/\partial\varphi$ is precisely $\bar{T}^{\varphi}_{\vartheta\varphi}\triangle\vartheta\triangle\varphi$. So, one gets (recalling that $\nabla_{\partial_j}\partial_i = \Gamma^k_{ji}\partial_k$)

$$\bar{T}^{\varphi}_{\vartheta\varphi} = \left(\Gamma^{\varphi}_{\vartheta\varphi} - \Gamma^{\varphi}_{\varphi\vartheta}\right) = -\cot\theta. \tag{B.16}$$

To end this Appendix, it is worth showing that $\nabla$ is metrically compatible, i.e., that $\nabla\mathbf{g} = 0$. Indeed, we have:

$$\begin{aligned} 0 = &\ \nabla_{\mathbf{e_c}}\mathbf{g}(\mathbf{e_i}, \mathbf{e_j}) = (\nabla_{\mathbf{e_c}}\mathbf{g})(\mathbf{e_i}, \mathbf{e_j}) + \mathbf{g}(\nabla_{\mathbf{e_c}}\mathbf{e_i}, \mathbf{e_j}) + \mathbf{g}(\mathbf{e_i}, \nabla_{\mathbf{e_c}}\mathbf{e_j}) \\ = &\ (\nabla_{\mathbf{e_c}}\mathbf{g})(\mathbf{e_i}, \mathbf{e_j}). \end{aligned} \tag{B.17}$$

[3]This wording, of course, means that those vectors are identified as elements of the appropriate tangent spaces.

# References

[1] Bergmann, P. G., *The Riddle of Gravitation*, Dover, New York, 1992.
[2] do Carmo, M. P., *Riemannian Geometry*, Birkhhäuser, Boston 1992.

## References

[illegible]

[illegible]

# Appendix C
## Gravitational Theory for Independent $\boldsymbol{h}$ and $\Omega$ Fields

In this Appendix, we will present, for completeness[1], a theory where the dynamics of the gravitational field is such that it must be described by two independent fields [1], namely the plastic distortion field $\boldsymbol{h}$ (more precisely $\boldsymbol{h}^{\clubsuit}$) and a $\varkappa\eta$-gauge rotation field $\Omega \equiv \overset{\varkappa\eta}{\Omega}$ defined on the canonical space $U$. These two independent fields distort the Lorentz vacuum, creating an effective Minkowski-Cartan *MCGSS* $(U, \eta, \varkappa)$, or equivalently a Lorentz-Cartan *MCGSS* $(U, \boldsymbol{g} = \boldsymbol{h}^{\dagger}\eta\boldsymbol{h}, \gamma)$ which is viewed as gauge equivalent to $(U, \eta, \varkappa)$ as already discussed in the main text. From the mathematical point of view, our theory is then described by a Lagrangian, where the fields $\boldsymbol{h}$ and $\Omega$ are the dynamic variables to be varied in the action functional. The Lagrangian is postulated to be the scalar functional

$$[\boldsymbol{h}^{\clubsuit}, \Omega, \cdot\partial\Omega] \mapsto \mathfrak{L}[\boldsymbol{h}^{\clubsuit}, \Omega, \cdot\partial\Omega] = \mathcal{R}\det[\boldsymbol{h}], \tag{C.1}$$

where $\mathcal{R}(= \overset{\gamma h}{\mathcal{R}})$ (given by Eq.(4.114) ) is expressed as a functional of the fields $\boldsymbol{h}^{\clubsuit}$, $\Omega$ and $\cdot\partial\Omega$, i.e.,

$$\begin{aligned} \mathcal{R} &= \underline{\boldsymbol{h}}^{\clubsuit}(\partial_a \wedge \partial_b) \cdot \mathcal{R}_2(a \wedge b) \\ &= \underline{\boldsymbol{h}}^{\clubsuit}(\partial_a \wedge \partial_b) \cdot (a \cdot \partial\Omega(b) - b \cdot \partial\Omega(a) - \Omega([a, b]) + \Omega(a) \underset{\eta}{\times} \Omega(b)), \end{aligned} \tag{C.2}$$

and $\det[\boldsymbol{h}]$, as in Section 5.1 is expressed as a functional of $\boldsymbol{h}^{\clubsuit}$, i.e., $\det[\boldsymbol{h}] = (\det[\boldsymbol{h}^{\clubsuit}])^{-1}$.

The Lagrangian formalism for extensor fields supposes the existence of an action functional

$$\mathcal{A} = \int_U \mathfrak{L}[\boldsymbol{h}^{\clubsuit}, \Omega, \cdot\partial\Omega]\ \tau = \int_U \mathcal{R}\det[\boldsymbol{h}]\ \tau \tag{C.3}$$

which according to the principle of stationary action, gives for any of the dynamic variables a contour problem.

---

[1] And also, as a justification for our construction in Section 6.4 of the spin angular momentum of the gravitational field in our theory.

For the dynamic variable $\boldsymbol{h}^{\clubsuit}$, it is

$$\int_U \delta^w_{h^\clubsuit} \mathfrak{L}[\boldsymbol{h}^{\clubsuit}, \Omega, \cdot\partial\Omega]\ \tau = 0, \tag{C.4}$$

for any smooth $(1,1)$-extensor field $w$ such that $w|_{\partial U} = 0$.

For the dynamic variable $\Omega$, it is

$$\int_U \delta^B_\Omega \mathfrak{L}[\boldsymbol{h}^{\clubsuit}, \Omega, \cdot\partial\Omega]\ \tau = 0, \tag{C.5}$$

for any smooth $(1,2)$-extensor field $B$, such that $B|_{\partial U} = 0$.

We now show that Eqs.(C.4) and (C.5) lead to coupled differential equations for the fields $\boldsymbol{h}^{\clubsuit}$ and $\Omega$.

Of course, the solution of the contour problems given by Eqs.(C.4) and (C.5) is obtained following the traditional path, i.e., by using the proper variational formulas, the Gauss-Stokes and a fundamental lemma of integration theory (already mentioned in Section 5.1).

First, we calculate the variation in Eq.(C.4). We have:

$$\begin{aligned}\boldsymbol{\delta}^w_{h^\clubsuit}(\underline{\boldsymbol{h}}^{\clubsuit}(\partial_a \wedge \partial_b) \cdot \mathcal{R}_2(a \wedge b) \det[\boldsymbol{h}]) &= (\boldsymbol{\delta}^w_{h^\clubsuit} \underline{\boldsymbol{h}}^{\clubsuit}(\partial_a \wedge \partial_b)) \cdot \mathcal{R}_2(a \wedge b) \det[\boldsymbol{h}] \\ &+ \underline{\boldsymbol{h}}^{\clubsuit}(\partial_a \wedge \partial_b) \cdot \mathcal{R}_2(a \wedge b)(\delta^w_{\boldsymbol{h}^\clubsuit} \det[\boldsymbol{h}]).\end{aligned} \tag{C.6}$$

For the scalar product in the first member of Eq.(C.6), an algebraic manipulation analogous to the one already used for deriving Eq.(5.15) gives

$$(\boldsymbol{\delta}^w_{h^\clubsuit} \underline{\boldsymbol{h}}^{\clubsuit}(\partial_a \wedge \partial_b)) \cdot \mathcal{R}_2(a \wedge b) = 2w(\partial_a) \cdot \mathcal{R}_1(a), \tag{C.7}$$

where $\mathcal{R}(a)$ is the Ricci gauge field (in the Lorentz-Cartan *MCGSS*).

The calculation of $\boldsymbol{\delta}^w_{h^\clubsuit} \det[\boldsymbol{h}]$ has already been obtained (see Eq.(5.22)). So, substituting Eqs.(C.7) and (5.22) in Eq.(C.6) and recalling the definition of $\mathcal{G} \equiv \overset{\gamma_h}{\mathcal{G}}$ (Eq.(4.115)), we get that

$$\begin{aligned}\boldsymbol{\delta}^w_{h^\clubsuit}(\mathcal{R} \det[\boldsymbol{h}]) &= 2w(\partial_a) \cdot \mathcal{R}_1(a) \det[\boldsymbol{h}] - \mathcal{R} w(\partial_a) \cdot \boldsymbol{h}(a) \det[\boldsymbol{h}] \\ &= 2w(\partial_a) \cdot (\mathcal{R}_1(a) - \frac{1}{2}\boldsymbol{h}(a)\mathcal{R}) \det[\boldsymbol{h}] \\ \boldsymbol{\delta}^w_{h^\clubsuit}(\mathcal{R} \det[\boldsymbol{h}]) &= 2w(\partial_a) \cdot \mathcal{G}(a) \det[\boldsymbol{h}].\end{aligned} \tag{C.8}$$

Substituting Eq.(C.8) in Eq.(C.4), the contour problem for the dynamic variable $\boldsymbol{h}^{\clubsuit}$ becomes

$$\int_{\mathcal{U}} w(\partial_a) \cdot \mathcal{G}(a) \det[\boldsymbol{h}]\ \tau = 0, \tag{C.9}$$

for any $w$, and since $w$ is arbitrary, recalling a fundamental lemma of integration theory, we get

$$\mathcal{G}(a) = 0. \tag{C.10}$$

This field equation looks like the one already obtained for $\boldsymbol{h}$ in Section 6.1, but here, it is not a differential equation for the field $\boldsymbol{h}$. To continue, recall that Eq.(C.10) is equivalent to $\mathcal{R}(a) = 0$.

Indeed, taking the scalar $\boldsymbol{h}^{\clubsuit}$-divergent of $\mathcal{R}_1(a)$, we have

$$\begin{aligned}\mathcal{G}(a) = 0 &\Rightarrow \mathcal{R}_1(a) - \frac{1}{2}\boldsymbol{h}(a)\mathcal{R} = 0\\ &\Rightarrow \boldsymbol{h}^{\clubsuit}(\partial_a) \cdot \mathcal{R}_1(a) - \frac{1}{2}\boldsymbol{h}^{\clubsuit}(\partial_a) \cdot \boldsymbol{h}(a)\mathcal{R} = 0\\ &\Rightarrow \mathcal{R} - \frac{1}{2}4\mathcal{R} = 0\\ &\Rightarrow \mathcal{R} = 0,\end{aligned}$$

which implies

$$\mathcal{R}_1(a) = 0. \tag{C.11}$$

Also, it is quite clear that Eq.(C.11) implies Eq.(C.10).

Let us express now Eq.(C.11) in terms of the *canonical* 1-forms of $\boldsymbol{h}$ (i.e., $\boldsymbol{h}(\boldsymbol{\vartheta}^0), \ldots, \boldsymbol{h}(\boldsymbol{\vartheta}^3)$) and the canonical *biforms* of $\Omega$ (i.e., $\Omega(\vartheta_0), \ldots, \Omega(\vartheta_3)$),

$$\begin{aligned}\mathcal{R}_1(a) = 0 \Leftrightarrow \mathcal{R}_1(\vartheta_\nu) &= 0\\ \Leftrightarrow \boldsymbol{h}^{\clubsuit}(\boldsymbol{\vartheta}^\mu) \lrcorner \mathcal{R}_2(\vartheta_\mu \wedge \vartheta_\nu) &= 0,\end{aligned}$$

i.e.,

$$\boldsymbol{h}^{\clubsuit}(\boldsymbol{\vartheta}^\mu) \lrcorner (\vartheta_\mu \cdot \partial\Omega(\vartheta_\nu) - \vartheta_\nu \cdot \partial\Omega(\vartheta_\mu) + \Omega(\vartheta_\mu) \underset{\eta}{\times} \Omega(\vartheta_\nu)) = 0. \tag{C.12}$$

Eq.(C.12) is a set of four coupled multiform equations which are algebraic on the canonical *1*-forms of $\boldsymbol{h}$ and differential on the canonical biforms of $\Omega$.

Let us, now, calculate the variation in Eq.(C.5). We have

$$\boldsymbol{\delta}_\Omega^B(\underline{\boldsymbol{h}}^{\clubsuit}(\partial_a \wedge \partial_b) \cdot \mathcal{R}_2(a \wedge b) \det[\boldsymbol{h}]) = \underline{\boldsymbol{h}}^{\clubsuit}(\partial_a \wedge \partial_b) \cdot (\boldsymbol{\delta}_\Omega^B \mathcal{R}_2(a \wedge b)) \det[\boldsymbol{h}]. \tag{C.13}$$

For the scalar product on the right side of Eq.(C.13), using the commutation property involving $\boldsymbol{\delta}_\Omega^B$ and $a \cdot \partial$, the property $\boldsymbol{\delta}_\Omega^B \Omega(a) = B(a)$ and the definition of the covariant derivative operator $\mathcal{D}_a \equiv \overset{\varkappa\eta}{\mathcal{D}}_a$, i.e., $\mathcal{D}_a X = a \cdot \partial X + \Omega(a) \underset{\eta}{\times} X$, we have

$$\begin{aligned}\underline{\boldsymbol{h}}^{\clubsuit}(\partial_a \wedge \partial_b) \cdot (\boldsymbol{\delta}_\Omega^B \mathcal{R}_2(a \wedge b)) &= \underline{\boldsymbol{h}}^{\clubsuit}(\boldsymbol{\vartheta}^\mu \wedge \boldsymbol{\vartheta}^\nu) \cdot (\boldsymbol{\delta}_\Omega^B \mathcal{R}_2(\vartheta_\mu \wedge \vartheta_\nu))\\ &= \underline{\boldsymbol{h}}^{\clubsuit}(\boldsymbol{\vartheta}^\mu \wedge \boldsymbol{\vartheta}^\nu) \cdot (\vartheta_\mu \cdot \partial B(\vartheta_\nu) - \vartheta_\nu \cdot \partial B(\vartheta_\mu)\\ &+ B(\vartheta_\mu) \underset{\eta^{-1}}{\times} \Omega(\vartheta_\nu) + \Omega(\vartheta_\mu) \underset{\eta^{-1}}{\times} B(\vartheta_\nu))\\ &= 2\underline{\boldsymbol{h}}^{\clubsuit}(\boldsymbol{\vartheta}^\mu \wedge \boldsymbol{\vartheta}^\nu) \cdot \mathcal{D}_{\vartheta_\mu} B(\vartheta_\nu).\end{aligned} \tag{C.14}$$

After some algebraic manipulations, the right side of Eq.(C.14) may be written as

$$\begin{aligned}
\underline{\boldsymbol{h}}^{\clubsuit}(\boldsymbol{\vartheta}^{\mu} \wedge \boldsymbol{\vartheta}^{\nu}) \cdot \mathcal{D}_{\vartheta_{\mu}} B(\vartheta_{\nu}) &= \underline{\boldsymbol{h}}^{\clubsuit}(\boldsymbol{\vartheta}^{\mu} \wedge \boldsymbol{\vartheta}^{\nu}) \cdot \vartheta_{\mu} \cdot \partial B(\vartheta_{\nu}) \\
&+ \vartheta_{\mu} \cdot \partial \underline{\boldsymbol{h}}^{\clubsuit}(\boldsymbol{\vartheta}^{\mu} \wedge \boldsymbol{\vartheta}^{\nu}) \cdot B(\vartheta_{\nu}) \\
&+ \underline{\boldsymbol{h}}^{\clubsuit}(\boldsymbol{\vartheta}^{\mu} \wedge \boldsymbol{\vartheta}^{\nu}) \cdot (\Omega(\vartheta_{\mu}) \underset{\eta^{-1}}{\times} B(\vartheta_{\nu})) \\
&- \vartheta_{\mu} \cdot \partial \underline{\boldsymbol{h}}^{\clubsuit}(\boldsymbol{\vartheta}^{\mu} \wedge \boldsymbol{\vartheta}^{\nu}) \cdot B(\vartheta_{\nu}) \\
&= \vartheta_{\mu} \cdot \partial(\underline{\boldsymbol{h}}^{\clubsuit}(\boldsymbol{\vartheta}^{\mu} \wedge \boldsymbol{\vartheta}^{\nu}) \cdot B(\vartheta_{\nu})) - (\vartheta_{\mu} \cdot \partial \underline{\boldsymbol{h}}^{\clubsuit}(\boldsymbol{\vartheta}^{\mu} \wedge \boldsymbol{\vartheta}^{\nu}) \\
&+ \underline{\eta}(\Omega(\vartheta_{\mu}) \underset{\eta}{\times} \underline{\eta}\underline{\boldsymbol{h}}^{\clubsuit}(\boldsymbol{\vartheta}^{\mu} \wedge \boldsymbol{\vartheta}^{\nu}))) \cdot B(\vartheta_{\nu}) \\
&= \vartheta_{\mu} \cdot \partial(\underline{\boldsymbol{h}}^{\clubsuit}(\boldsymbol{\vartheta}^{\mu} \wedge \boldsymbol{\vartheta}^{\nu}) \cdot B(\vartheta_{\nu})) - \underline{\eta}(\vartheta_{\mu} \cdot \partial \underline{\eta}\underline{\boldsymbol{h}}^{\clubsuit}(\boldsymbol{\vartheta}^{\mu} \wedge \vartheta^{\nu}) \\
&+ \Omega(\vartheta_{\mu}) \underset{\eta}{\times} \underline{\eta}\underline{\boldsymbol{h}}^{\clubsuit}(\boldsymbol{\vartheta}^{\mu} \wedge \boldsymbol{\vartheta}^{\nu})) \cdot B(\vartheta_{\nu}),
\end{aligned}$$

i.e.,

$$\begin{aligned}
\underline{\boldsymbol{h}}^{\clubsuit}(\boldsymbol{\vartheta}^{\mu} \wedge \boldsymbol{\vartheta}^{\nu}) \cdot \mathcal{D}_{\vartheta_{\mu}} B(\vartheta_{\nu}) &= \vartheta_{\mu} \cdot \partial(\underline{\boldsymbol{h}}^{\clubsuit}(\boldsymbol{\vartheta}^{\mu} \wedge \boldsymbol{\vartheta}^{\nu}) \cdot B(\vartheta_{\nu})) \\
&- \underline{\eta}\mathcal{D}_{\vartheta_{\mu}} \underline{\eta}\underline{\boldsymbol{h}}^{\clubsuit}(\boldsymbol{\vartheta}^{\mu} \wedge \boldsymbol{\vartheta}^{\nu}) \cdot B(\vartheta_{\nu}).
\end{aligned} \tag{C.15}$$

Also, the first member on the right side of Eq.(C.15) may be written as a scalar divergent of a 1-form field, i.e.,

$$\vartheta_{\mu} \cdot \partial(\underline{\boldsymbol{h}}^{\clubsuit}(\boldsymbol{\vartheta}^{\mu} \wedge \boldsymbol{\vartheta}^{\nu}) \cdot B(\vartheta_{\nu})) = \partial \cdot (\vartheta_{\mu}\underline{\boldsymbol{h}}^{\clubsuit}(\boldsymbol{\vartheta}^{\mu} \wedge \boldsymbol{\vartheta}^{\nu}) \cdot B(\vartheta_{\nu})), \tag{C.16}$$

where we used the fact that $\partial \cdot (\vartheta_{\mu}\phi) = \vartheta_{\mu} \cdot \partial\phi$, where $\phi$ is a scalar field.

After putting Eq.(C.16) in Eq.(C.15) and using the result obtained in Eq.(C.14), the Eq.(C.13) gives us finally the variational formula

$$\begin{aligned}
\boldsymbol{\delta}_{\Omega}^{B}(\underline{\boldsymbol{h}}^{\clubsuit}(\partial_a \wedge \partial_b) \cdot \mathcal{R}_2(a \wedge b)\det[\boldsymbol{h}]) &= -2(\underline{\eta}\mathcal{D}_{\vartheta_{\mu}}\underline{\eta}\underline{\boldsymbol{h}}^{\clubsuit}(\boldsymbol{\vartheta}^{\mu} \wedge \boldsymbol{\vartheta}^{\nu}) \cdot B(\vartheta_{\nu}) \\
&- \partial \cdot (\vartheta_{\mu}\underline{\boldsymbol{h}}^{\clubsuit}(\boldsymbol{\vartheta}^{\mu} \wedge \boldsymbol{\vartheta}^{\nu}) \cdot B(\vartheta_{\nu})))\det[\boldsymbol{h}].
\end{aligned} \tag{C.17}$$

Putting Eq.(C.17) in Eq.(C.5), the contour problem for the dynamic variable $\Omega$ becomes

$$\begin{aligned}
&\int_U \underline{\eta}\mathcal{D}_{\vartheta_{\mu}}\underline{\eta}\underline{\boldsymbol{h}}^{\clubsuit}(\boldsymbol{\vartheta}^{\mu} \wedge \boldsymbol{\vartheta}^{\nu}) \cdot B(\vartheta_{\nu})\det[\boldsymbol{h}]\ \tau \\
&- \int_U \partial \cdot (\vartheta_{\mu}\underline{\boldsymbol{h}}^{\clubsuit}(\boldsymbol{\vartheta}^{\mu} \wedge \boldsymbol{\vartheta}^{\nu}) \cdot B(\vartheta_{\nu}))\det[\boldsymbol{h}]\ \tau = 0,
\end{aligned} \tag{C.18}$$

for all $B$ such that $B|_{\partial U} = 0$. Now, the second member of Eq.(C.18) may be easily transformed in such a way that a scalar divergent of a 1-form field appears. Indeed, if we recall that and $\partial \cdot (a\phi) = a \cdot \partial\phi + (\partial \cdot a)\phi$, where $a$ is a 1-form field and $\phi$ is a scalar field, we have

$$\begin{aligned}
&\int_U \partial \cdot (\vartheta_{\mu}\underline{\boldsymbol{h}}^{\clubsuit}(\boldsymbol{\vartheta}^{\mu} \wedge \boldsymbol{\vartheta}^{\nu}) \cdot B(\vartheta_{\nu}))\det[\boldsymbol{h}]\ \tau \\
&= \int_U \partial \cdot (\vartheta_{\mu}\underline{\boldsymbol{h}}^{\clubsuit}(\boldsymbol{\vartheta}^{\mu} \wedge \boldsymbol{\vartheta}^{\nu}) \cdot B(\vartheta_{\nu}))\det[\boldsymbol{h}])\ \tau \\
&- \int_U \underline{\boldsymbol{h}}^{\clubsuit}(\boldsymbol{\vartheta}^{\mu} \wedge \boldsymbol{\vartheta}^{\nu}) \cdot B(\vartheta_{\nu})\vartheta_{\mu} \cdot \partial\det[\boldsymbol{h}]\ \tau.
\end{aligned} \tag{C.19}$$

Putting Eq.(C.19) in Eq.(C.18), we get

$$\int_U (\underline{\eta} \mathcal{D}_{\vartheta_\mu} \underline{\eta} \underline{\boldsymbol{h}}^\clubsuit(\boldsymbol{\vartheta}^\mu \wedge \boldsymbol{\vartheta}^\nu) \det[\boldsymbol{h}] + \underline{\boldsymbol{h}}^\clubsuit(\boldsymbol{\vartheta}^\mu \wedge \boldsymbol{\vartheta}^\nu)\vartheta_\mu \cdot \partial \det[\boldsymbol{h}]) \cdot B(\vartheta_\nu)\, \tau$$
$$- \int_U \partial \cdot (\vartheta_\mu \underline{\boldsymbol{h}}^\clubsuit(\boldsymbol{\vartheta}^\mu \wedge \boldsymbol{\vartheta}^\nu) \cdot B(\vartheta_\nu)) \det[\boldsymbol{h}])\, \tau = 0, \tag{C.20}$$

for all $B$ such that $B|_{\partial U} = 0$. Next, using the Gauss-Stokes theorem with the boundary condition $B|_{\partial U} = 0$, the second term of Eq.(C.20) may be integrated and gives

$$\begin{aligned}
&\int_U \partial \cdot (\vartheta_\mu \underline{\boldsymbol{h}}^\clubsuit(\boldsymbol{\vartheta}^\mu \wedge \boldsymbol{\vartheta}^\nu) \cdot B(\boldsymbol{\vartheta}_\nu)) \det[\boldsymbol{h}])\, \tau \\
&= \int_{\partial U} \boldsymbol{\vartheta}^\rho \cdot \vartheta_\mu \underline{\boldsymbol{h}}^\clubsuit(\boldsymbol{\vartheta}^\mu \wedge \boldsymbol{\vartheta}^\nu) \cdot B(\vartheta_\nu) \det[\boldsymbol{h}]\, \tau_\rho \\
&= \int_{\partial U} \underline{\boldsymbol{h}}^\clubsuit(\boldsymbol{\vartheta}^\mu \wedge \boldsymbol{\vartheta}^\nu) \cdot B(\vartheta_\nu) \det[\boldsymbol{h}]\, \tau_\mu = 0.
\end{aligned} \tag{C.21}$$

So, putting Eq.(C.21) in Eq.(C.20) gives

$$\int_U (\underline{\eta} \mathcal{D}_{\vartheta_\mu} \underline{\eta} \underline{\boldsymbol{h}}^\clubsuit(\boldsymbol{\vartheta}^\mu \wedge \boldsymbol{\vartheta}^\nu) \det[\boldsymbol{h}] + \underline{\boldsymbol{h}}^\clubsuit(\boldsymbol{\vartheta}^\mu \wedge \boldsymbol{\vartheta}^\nu)\vartheta_\mu \cdot \partial \det[\boldsymbol{h}]) \cdot B(\vartheta_\nu)\, \tau = 0, \tag{C.22}$$

for all $B$ such that $B|_{\partial U} = 0$, and since $B$ is arbitrary a fundamental lemma of integration theory finally yields

$$\underline{\eta} \mathcal{D}_{\vartheta_\mu} \underline{\eta} \underline{\boldsymbol{h}}^\clubsuit(\boldsymbol{\vartheta}^\mu \wedge \boldsymbol{\vartheta}^\nu) \det[\boldsymbol{h}] + \underline{\boldsymbol{h}}^\clubsuit(\boldsymbol{\vartheta}^\mu \wedge \boldsymbol{\vartheta}^\nu)\vartheta_\mu \cdot \partial \det[\boldsymbol{h}] = 0,$$

i.e.,

$$\mathcal{D}_{\vartheta_\mu} \underline{\eta} \underline{\boldsymbol{h}}^\clubsuit(\boldsymbol{\vartheta}^\mu \wedge \boldsymbol{\vartheta}^\nu) = -(\vartheta_\mu \cdot \partial \ln |\det[\boldsymbol{h}]|) \underline{\eta} \underline{\boldsymbol{h}}^*(\boldsymbol{\vartheta}^\mu \wedge \boldsymbol{\vartheta}^\nu), \tag{C.23}$$

or yet,

$$(\vartheta_\mu \cdot \partial \ln |\det[\boldsymbol{h}]|) \underline{\eta} \underline{\boldsymbol{h}}^\clubsuit(\boldsymbol{\vartheta}^\mu \wedge \boldsymbol{\vartheta}^\nu) + \vartheta_\mu \cdot \partial \underline{\eta} \underline{\boldsymbol{h}}^\clubsuit(\boldsymbol{\vartheta}^\mu \wedge \boldsymbol{\vartheta}^\nu) + \Omega(\vartheta_\mu) \underset{\eta}{\times} \underline{\eta} \underline{\boldsymbol{h}}^\clubsuit(\boldsymbol{\vartheta}^\mu \wedge \boldsymbol{\vartheta}^\nu) = 0. \tag{C.24}$$

This is the field equation for the $\varkappa\eta$-gauge rotation field $\Omega$. It is a set of four coupled multiform equations, differential in the canonical 1-forms of $\boldsymbol{h}^\clubsuit$ (i.e., $\boldsymbol{h}^\clubsuit(\boldsymbol{\vartheta}^0), \ldots, \boldsymbol{h}^\clubsuit(\boldsymbol{\vartheta}^3)$), and algebraic in the canonical biforms of $\Omega$ (i.e., $\Omega(\vartheta_0), \ldots, \Omega(\vartheta_3)$).

We end this Appendix, writing again the Euler-Lagrange equations for the fields $\boldsymbol{h}$ and $\Omega$, in terms of the canonical 1-form and biform fields.

$$\boldsymbol{h}^\clubsuit(\boldsymbol{\vartheta}^\mu) \lrcorner (\vartheta_\mu \cdot \partial\Omega(\vartheta_\nu) - \vartheta_\nu \cdot \partial\Omega(\vartheta_\mu) + \Omega(\vartheta_\mu) \underset{\eta}{\times} \Omega(\vartheta_\nu)) = 0, \tag{C.25}$$

$$(\vartheta_\mu \cdot \partial \ln |\det[\boldsymbol{h}]|) \underline{\eta} \underline{\boldsymbol{h}}^\clubsuit(\boldsymbol{\vartheta}^\mu \wedge \boldsymbol{\vartheta}^\nu) + \vartheta_\mu \cdot \partial \underline{\eta} \underline{\boldsymbol{h}}^\clubsuit(\boldsymbol{\vartheta}^\mu \wedge \boldsymbol{\vartheta}^\nu) + \Omega(\vartheta_\mu) \underset{\eta}{\times} \underline{\eta} \underline{\boldsymbol{h}}^\clubsuit(\boldsymbol{\vartheta}^\mu \wedge \boldsymbol{\vartheta}^\nu) = 0. \tag{C.26}$$

# References

Fernández, V. V., Distortion and Rotation Extensors in Einstein's Gravitational Theory, Ph.D. thesis (in Portuguese), IFGW-UNICAMP, 2000.

# Appendix D
## Proof of Eq.(6.13)

**Proposition** [1] The Einstein-Hilbert Lagrangian density $L_{eh}$ can be written as

$$\mathcal{L}_{eh} = -d\left(\mathfrak{g}^{\alpha} \wedge \underset{g}{\star} d\mathfrak{g}_{\alpha}\right) + \mathcal{L}_g, \tag{D.1}$$

where

$$\mathcal{L}_g = -\frac{1}{2} d\mathfrak{g}^{\alpha} \wedge \underset{g}{\star} d\mathfrak{g}_{\alpha} + \frac{1}{2} \underset{g}{\delta}\mathfrak{g}^{\alpha} \wedge \underset{g}{\star}\underset{g}{\delta}\mathfrak{g}_{\alpha} + \frac{1}{4} d\mathfrak{g}^{\alpha} \wedge \mathfrak{g}_{\alpha} \wedge \underset{g}{\star}(d\mathfrak{g}^{\alpha} \wedge \mathfrak{g}_{\alpha}) \tag{D.2}$$

is the first order Lagrangian density (first introduced by Einstein).

**Proof**: We will establish that $\mathcal{L}_g$ may be written as

$$\mathcal{L}_g = -\frac{1}{2}\tau_g \mathfrak{g}^{\gamma} \underset{g^{-1}}{\lrcorner} \mathfrak{g}^{\beta} \underset{g^{-1}}{\lrcorner} \left(\omega_{\alpha\beta} \wedge \omega_{\gamma}^{\alpha}\right), \tag{D.3}$$

which is nothing more than the *intrinsic* form of the Einstein first order Lagrangian, in the gauge defined by the basis $\{\mathfrak{g}^{\alpha}\}$. To see this first recall Cartan's structure equations for the Lorentzian structure $(M \simeq \mathbb{R}^4, \mathbf{g}, D)$, i.e.,

$$d\mathfrak{g}^{\alpha} = -\omega_{\beta}^{\alpha} \wedge \mathfrak{g}^{\beta}, \tag{D.4a}$$

$$\mathcal{R}_{\beta}^{\alpha} = d\omega_{\beta}^{\alpha} + \omega_{\gamma}^{\alpha} \wedge \omega_{\beta}^{\gamma}, \tag{D.4b}$$

where $\omega_{\mathbf{b}}^{\mathbf{a}}$ are the so called connection 1-form fields, which on the basis $\{\mathfrak{g}^{\mathbf{a}}\}$ are written as

$$\omega_{\beta}^{\alpha} = L_{\beta\gamma}^{\alpha} \mathfrak{g}^{\gamma}. \tag{D.5}$$

Now, using one of the identities in Eq.(2.84), we easily get

$$\mathfrak{g}^{\gamma} \underset{g^{-1}}{\lrcorner} \mathfrak{g}^{\beta} \underset{g^{-1}}{\lrcorner} \left(\omega_{\alpha\beta} \wedge \omega_{\gamma}^{\alpha}\right) = \eta^{\beta\kappa}\left(L_{\kappa\gamma}^{\delta} L_{\delta\beta}^{\gamma} - L_{\delta\gamma}^{\delta} L_{\kappa\beta}^{\gamma}\right). \tag{D.6}$$

Next, we recall that Eq.(D.4a) can easily be solved for $\omega_{\mathfrak{b}}^{\mathfrak{a}}$ in terms of the $d\mathfrak{g}^{\mathbf{a}}$'s and the $\mathfrak{g}^{\mathbf{b}}$'s. We get

$$\omega^{\gamma\delta} = \frac{1}{2}\left[\mathfrak{g}^{\delta} \underset{g^{-1}}{\lrcorner} d\mathfrak{g}^{\gamma} - \mathfrak{g}^{\gamma} \underset{g^{-1}}{\lrcorner} d\mathfrak{g}^{\delta} + \mathfrak{g}^{\gamma} \underset{g^{-1}}{\lrcorner} \left(\mathfrak{g}^{\delta} \underset{g^{-1}}{\lrcorner} d\mathfrak{g}_{\alpha}\right) \mathfrak{g}^{\alpha}\right]. \tag{D.7}$$

We now have all ingredients to prove that $\mathcal{L}_{eh}$ can be written as in Eq.(D.1). Indeed, using Cartan's second structure equation (Eq.(D.4b)), we can write Eq.(6.11) as:

$$\begin{aligned}
\mathcal{L}_{eh} &= \frac{1}{2} d\omega_{\alpha\beta} \wedge \underset{g}{\star}(\mathfrak{g}^{\alpha} \wedge \mathfrak{g}^{\beta}) + \frac{1}{2}\omega_{\alpha\gamma} \wedge \omega_{\beta}^{\gamma} \wedge \underset{g}{\star}(\mathfrak{g}^{\alpha} \wedge \mathfrak{g}^{\beta}) \\
&= \frac{1}{2} d[\omega_{\alpha\beta} \wedge \underset{g}{\star}(\mathfrak{g}^{\alpha} \wedge \mathfrak{g}^{\beta})] + \frac{1}{2}\omega_{\alpha\beta} \wedge \underset{g}{\star} d(\mathfrak{g}^{\alpha} \wedge \mathfrak{g}^{\beta}) + \frac{1}{2}\omega_{\alpha\gamma} \wedge \omega_{\beta}^{\gamma} \wedge \underset{g}{\star}(\mathfrak{g}^{\alpha} \wedge \mathfrak{g}^{\beta}) \\
&= \frac{1}{2} d[\omega_{\alpha\beta} \wedge \underset{g}{\star}(\mathfrak{g}^{\alpha} \wedge \mathfrak{g}^{\beta})] - \frac{1}{2}\omega_{\alpha\beta} \wedge \omega_{\gamma}^{\alpha} \wedge \underset{g}{\star}(\mathfrak{g}^{\gamma} \wedge \mathfrak{g}^{\beta}).
\end{aligned} \tag{D.8}$$

Next, using again some of the identities in Eq.(2.102) we get (recall that $\omega^{\gamma\delta} = -\omega^{\delta\gamma}$)

$$\begin{aligned}
(\mathfrak{g}^{\gamma} \wedge \mathfrak{g}^{\delta}) \wedge \underset{g}{\star}\omega_{\gamma\delta} &= -\underset{g}{\star}\left[\omega^{\gamma\delta} \underset{g}{\lrcorner} (\mathfrak{g}_{\gamma} \wedge \mathfrak{g}_{\delta})\right] \\
&= \underset{g}{\star}\left[(\omega^{\gamma\delta} \underset{g}{\cdot} \mathfrak{g}_{\delta})\theta_{\gamma} - (\omega^{\gamma\delta} \underset{g}{\cdot} \mathfrak{g}_{\gamma})\mathfrak{g}_{\delta}\right] = 2 \underset{g}{\star}\left[(\omega^{\gamma\delta} \underset{g}{\cdot} \mathfrak{g}_{\delta})\mathfrak{g}_{\gamma}\right],
\end{aligned} \tag{D.9}$$

and from Cartan's first structure equation (Eq.(D.4a)), we have

$$\begin{aligned}
\mathfrak{g}^{\alpha} \wedge \underset{g}{\star} d\mathfrak{g}_{\alpha} &= \mathfrak{g}^{\alpha} \wedge \underset{g}{\star}(\omega_{\beta\alpha} \wedge \mathfrak{g}^{\beta}) = -\underset{g}{\star}\left[\mathfrak{g}_{\alpha} \underset{g}{\lrcorner} (\omega^{\beta\alpha} \wedge \mathfrak{g}_{\beta})\right] \\
&= -\underset{g}{\star}\left[\mathfrak{g}_{\alpha} \underset{g}{\cdot} \omega^{\beta\alpha})\mathfrak{g}_{\beta}\right] = -\underset{g}{\star}\left[(\mathfrak{g}^{\alpha} \underset{g}{\cdot} \omega_{\beta\alpha})\mathfrak{g}^{\beta}\right],
\end{aligned} \tag{D.10}$$

from where it follows that

$$\frac{1}{2} d[(\mathfrak{g}^{\gamma} \wedge \mathfrak{g}^{\delta}) \wedge \underset{g}{\star}\omega_{\gamma\delta}] = -d(\mathfrak{g}^{\alpha} \wedge \underset{g}{\star} d\mathfrak{g}_{\alpha}). \tag{D.11}$$

On the other hand, the second term in the last line of Eq.(D.8) can be written as

$$\begin{aligned}
&\frac{1}{2}\omega_{\alpha\beta} \wedge \omega_{\gamma}^{\alpha} \wedge \underset{g}{\star}(\mathfrak{g}^{\gamma} \wedge \mathfrak{g}^{\beta}) \\
&= -\frac{1}{2} \underset{g}{\star}\left[(\mathfrak{g}^{\beta} \underset{g}{\cdot} \omega_{\alpha\beta})(\mathfrak{g}^{\gamma} \underset{g}{\cdot} \omega_{\gamma}^{\alpha}) - (\mathfrak{g}^{\beta} \underset{g}{\cdot} \omega_{\gamma}^{\alpha})(\mathfrak{g}^{\gamma} \underset{g}{\cdot} \omega_{\alpha\beta})\right].
\end{aligned}$$

Now,

$$\begin{aligned}
&(\mathfrak{g}^{\beta} \underset{g}{\cdot} \omega_{\alpha\beta})(\mathfrak{g}^{\gamma} \underset{g}{\cdot} \omega_{\gamma}^{\alpha}) \\
&= \omega_{\alpha\beta} \underset{g}{\cdot} \left[(\mathfrak{g}^{\gamma} \underset{g}{\cdot} \omega_{\gamma}^{\alpha})\mathfrak{g}^{\beta}\right] \\
&= \omega_{\alpha\beta} \underset{g}{\cdot} \left[\mathfrak{g}^{\gamma} \underset{g}{\lrcorner} (\omega_{\gamma}^{\alpha} \wedge \mathfrak{g}^{\beta}) + \omega^{\alpha\beta}\right] \\
&= (\omega_{\alpha\beta} \wedge \mathfrak{g}^{\gamma}) \underset{g}{\lrcorner} (\omega_{\gamma}^{\alpha} \wedge \mathfrak{g}^{\beta}) + \omega_{\gamma\delta} \underset{g}{\cdot} \omega^{\gamma\delta}\right]
\end{aligned}$$

and taking into account that $d\mathfrak{g}^{\alpha} = -\omega^{\alpha}_{\beta} \wedge \mathfrak{g}^{\beta}$, $\underset{g}{\star}^{-1} d \underset{g}{\star} \mathfrak{g}^{\alpha} = -\omega^{\alpha}_{\beta} \underset{g}{\cdot} \mathfrak{g}^{\beta}$, and that $\underset{g}{\delta} \mathfrak{g}_{\alpha} = -\underset{g}{\star}^{-1} d \underset{g}{\star} \mathfrak{g}_{\alpha}$, we have

$$\frac{1}{2}\omega_{\alpha\beta} \wedge \omega^{\alpha}_{\gamma} \wedge \underset{g}{\star}(\mathfrak{g}^{\gamma} \wedge \mathfrak{g}^{\beta}) = \frac{1}{2}[-d\mathfrak{g}^{\alpha} \wedge \underset{g}{\star} d\mathfrak{g}_{\alpha} - \underset{g}{\delta}\mathfrak{g}^{\alpha} \wedge \underset{g}{\star}\underset{g}{\delta}\mathfrak{g}_{\alpha} + \omega_{\gamma\delta} \wedge \underset{g}{\star}\omega^{\gamma\delta}]. \quad \text{(D.12)}$$

Next, using Eq.(D.7) the last term in the last equation can be written, after some algebra, as

$$\frac{1}{2}\omega_{\gamma\delta} \wedge \underset{g}{\star}\omega^{\gamma\delta} = d\mathfrak{g}^{\alpha} \wedge \underset{g}{\star} d\mathfrak{g}_{\alpha} - \frac{1}{4}\left(d\mathfrak{g}^{\alpha} \wedge \mathfrak{g}_{\alpha}\right) \wedge \underset{g}{\star}\left(d\mathfrak{g}^{\alpha} \wedge \mathfrak{g}_{\alpha}\right)$$

and we finally get

$$\mathcal{L}_g = -\frac{1}{2} d\mathfrak{g}^{\alpha} \wedge \underset{g}{\star} d\mathfrak{g}_{\alpha} + \frac{1}{2}\underset{g}{\delta}\mathfrak{g}^{\alpha} \wedge \underset{g}{\star}\underset{g}{\delta}\mathfrak{g}_{\alpha} + \frac{1}{4}\left(d\mathfrak{g}^{\alpha} \wedge \mathfrak{g}_{\alpha}\right) \wedge \underset{g}{\star}\left(d\mathfrak{g}^{\beta} \wedge \mathfrak{g}_{\beta}\right), \quad \text{(D.13)}$$

and the proposition is proved.

# References

[1] Rodrigues, W. A. Jr., and Capelas de Oliveira, E., T*he Many Faces of Maxwell, Dirac and Einstein Equations. A Clifford Bundle Approach.* Lecture Notes in Physics **722**, Springer, Heidelberg, 2007.

# Appendix E

# Derivation of the Field Equations from $\mathcal{L}_{eh}$

Let $X \in \sec \bigwedge^p T^*U$. *A multiform functional* $F$ of $X$ (not depending *explicitly* on $x \in \bigwedge^1 U$) is a mapping

$$F : \sec \bigwedge^p T^*U \to \sec \bigwedge^r T^*U$$

As, in Section 3, when no confusion arises, we use a sloppy notation and denote the image $F(X) \in \sec \bigwedge^r T^*U$ simply by $F$. Eventually, we also denote a functional $F$ by $F(X)$. Which object we are talking about is always obvious from the context of the equations where they appear.

Let $w := \boldsymbol{\delta} X \in \sec \bigwedge^p T^*U$. As in Exercise 3.6. we write the *variation* of $F$ in the direction of $\boldsymbol{\delta} X$ as the functional $\boldsymbol{\delta} F : \bigwedge^p U \to \bigwedge^r U$ given by

$$\boldsymbol{\delta} F = \lim_{\lambda \to 0} \frac{F(X + \lambda \boldsymbol{\delta} X) - F(X)}{\lambda}. \tag{E.1}$$

Recall from Remark 3.1 that the *algebraic derivative* of $F$ relative to $X$, $\frac{\partial F}{\partial X}$ is such that:

$$\boldsymbol{\delta} F = \boldsymbol{\delta} X \wedge \frac{\partial F}{\partial X}. \tag{E.2}$$

Moreover, given the functionals $F : \bigwedge^p U \to \bigwedge^r U \to$ and $G : \bigwedge^p U \to \bigwedge^s U$ the variation $\boldsymbol{\delta}$ satisfies

$$\boldsymbol{\delta}(F \wedge G) = \boldsymbol{\delta} F \wedge G + F \wedge \boldsymbol{\delta} G, \tag{E.3}$$

and the algebraic derivative (as is trivial to verify) satisfies

$$\frac{\partial}{\partial X}(F \wedge G) = \frac{\partial F}{\partial X} \wedge G + (-1)^{rp} F \wedge \frac{\partial G}{\partial X}. \tag{E.4}$$

An important property of $\boldsymbol{\delta}$ is that it commutes with the exterior derivative operator $d$, i.e., for any given functional $F$

$$d\boldsymbol{\delta} F = \boldsymbol{\delta} dF. \tag{E.5}$$

In general, we may have functionals depending on several different multiform fields, say, $F : \sec(\bigwedge^p T^*U \times \bigwedge^q T^*U) \to \sec \bigwedge^r T^*U$, with $(X, Y) \mapsto F(X, Y) \in \sec \bigwedge^r T^*U$. In this case, we have:

$$\boldsymbol{\delta} F = \boldsymbol{\delta} X \wedge \frac{\partial F}{\partial X} + \boldsymbol{\delta} Y \wedge \frac{\partial F}{\partial Y}. \tag{E.6}$$

An important case is the one where the functional $F$ is such that $F(X, dX) : \sec(\bigwedge^p T^*U \times \bigwedge^{p+1} T^*U) \to \sec \bigwedge^4 T^*U$. We then can write, supposing that the variation $\boldsymbol{\delta} X$ is chosen to be null on the boundary $\partial U'$, $U' \subset U$ (or that $\frac{\partial F}{\partial dX}\big|_{\partial U'} = 0$) and taking into account Stokes theorem,

$$\begin{aligned}
\boldsymbol{\delta} \int_{U'} F &= \int_{U'} \boldsymbol{\delta} F = \int_{U'} \boldsymbol{\delta} X \wedge \frac{\partial F}{\partial X} + \boldsymbol{\delta} dX \wedge \frac{\partial F}{\partial dX} \\
&= \int_{U'} \boldsymbol{\delta} X \wedge \left[\frac{\partial F}{\partial X} - (-1)^p d\left(\frac{\partial F}{\partial dX}\right)\right] + d\left(\boldsymbol{\delta} X \wedge \frac{\partial F}{\partial dX}\right) \\
&= \int_{U'} \boldsymbol{\delta} X \wedge \left[\frac{\partial F}{\partial X} - (-1)^p d\left(\frac{\partial F}{\partial dX}\right)\right] + \int_{\partial U'} \boldsymbol{\delta} X \wedge \frac{\partial F}{\partial dX} \\
&= \int_{U'} \boldsymbol{\delta} X \wedge \frac{\boldsymbol{\delta} F}{\boldsymbol{\delta} X},
\end{aligned} \tag{E.7}$$

where $\frac{\boldsymbol{\delta}}{\boldsymbol{\delta} X} F(X, dX) : \sec(\bigwedge^p T^*U \times \bigwedge^{p+1} T^*U) \to \bigwedge^{4-p} T^*U$ is called the *functional derivative* of $F$, and we have:

$$\frac{\boldsymbol{\delta} F}{\boldsymbol{\delta} X} = \frac{\partial F}{\partial X} - (-1)^p d\left(\frac{\partial F}{\partial dX}\right). \tag{E.8}$$

When $F = \mathcal{L}$ is a Lagrangian density in field theory $\frac{\boldsymbol{\delta}\mathcal{L}}{\boldsymbol{\delta} X}$ is the *Euler-Lagrange functional.*

We now obtain the variation of the Einstein-Hilbert Lagrangian density $\mathcal{L}_{eh}$, given by Eq.(6.11), i.e.,

$$\mathcal{L}_{eh} = \frac{1}{2}(\mathfrak{g}^\kappa \wedge \mathfrak{g}^\iota) \wedge \underset{g}{\star} \mathcal{R}_{\kappa\iota} = \frac{1}{2} \mathcal{R}_{\kappa\iota} \wedge \underset{g}{\star} (\mathfrak{g}^\kappa \wedge \mathfrak{g}^\iota),$$

by varying the $\omega_{\gamma\kappa}$ and the $\mathfrak{g}^\kappa$ independently.

We have[1]

$$\begin{aligned}
\boldsymbol{\delta} \mathcal{L}_{eh} &= \frac{1}{2} \boldsymbol{\delta} [\mathcal{R}_{\gamma\delta} \wedge \underset{g}{\star} (\mathfrak{g}^\gamma \wedge \mathfrak{g}^\delta)] \\
&= \frac{1}{2} \boldsymbol{\delta} \mathcal{R}_{\gamma\delta} \wedge \underset{g}{\star} (\mathfrak{g}^\gamma \wedge \mathfrak{g}^\delta) + \frac{1}{2} \mathcal{R}_{\gamma\delta} \wedge \boldsymbol{\delta} \underset{g}{\star} (\mathfrak{g}^\gamma \wedge \mathfrak{g}^\delta).
\end{aligned} \tag{E.9}$$

[1] We use only constrained variations of the $\mathfrak{g}^\alpha$ that do no change the metric field $\boldsymbol{g}$.

From Cartan's second structure equation, we can write

$$\begin{aligned}
&\boldsymbol{\delta}\mathcal{R}_{\gamma\delta}\wedge\underset{g}{\star}(\mathfrak{g}^{\gamma}\wedge\mathfrak{g}^{\delta})\\
&=\boldsymbol{\delta}d\omega_{\gamma\delta}\wedge\underset{g}{\star}(\mathfrak{g}^{\gamma}\wedge\mathfrak{g}^{\delta})+\boldsymbol{\delta}\omega_{\gamma\kappa}\wedge\omega_{\delta}^{\kappa}\wedge\star(\mathfrak{g}^{\gamma}\wedge\mathfrak{g}^{\delta})+\omega_{\gamma\kappa}\wedge\boldsymbol{\delta}\omega_{\delta}^{\kappa}\wedge\underset{g}{\star}(\mathfrak{g}^{\gamma}\wedge\mathfrak{g}^{\delta})\\
&=\boldsymbol{\delta}d\omega_{\gamma\delta}\wedge\underset{g}{\star}(\mathfrak{g}^{\gamma}\wedge\mathfrak{g}^{\delta}) \qquad\qquad (E.10)\\
&=d[\boldsymbol{\delta}\omega_{\gamma\delta}\wedge\underset{g}{\star}(\mathfrak{g}^{\gamma}\wedge\mathfrak{g}^{\delta})]-\boldsymbol{\delta}\omega_{\gamma\delta}\wedge d[\underset{g}{\star}(\mathfrak{g}^{\gamma}\wedge\mathfrak{g}^{\delta})].\\
&=d[\boldsymbol{\delta}\omega_{\gamma\delta}\wedge\underset{g}{\star}(\mathfrak{g}^{\gamma}\wedge\mathfrak{g}^{\delta})]-\boldsymbol{\delta}\omega_{\gamma\delta}\wedge[-\omega_{\kappa}^{\alpha}\wedge\underset{g}{\star}(\mathfrak{g}^{\kappa}\wedge\mathfrak{g}^{\kappa})-\omega_{\kappa}^{\delta}\wedge\underset{g}{\star}(\mathfrak{g}^{\gamma}\wedge\mathfrak{g}^{\kappa})]\\
&=d[\boldsymbol{\delta}\omega_{\gamma\delta}\wedge\underset{g}{\star}(\mathfrak{g}^{\gamma}\wedge\mathfrak{g}^{\delta})].
\end{aligned}$$

Moreover, using the definition of algebraic derivative (Eq.(E.1)), we have immediately

$$\boldsymbol{\delta}\underset{g}{\star}(\mathfrak{g}^{\gamma}\wedge\mathfrak{g}^{\delta})=\boldsymbol{\delta}\mathfrak{g}^{\kappa}\wedge\frac{\partial[\underset{g}{\star}(\mathfrak{g}^{\gamma}\wedge\mathfrak{g}^{\delta})]}{\partial\mathfrak{g}^{\kappa}} \qquad (E.11)$$

Now, recalling Eq.(2.103) of Chapter 4, we can write

$$\begin{aligned}
\boldsymbol{\delta}\underset{g}{\star}(\mathfrak{g}^{\gamma}\wedge\mathfrak{g}^{\delta})&=\boldsymbol{\delta}(\frac{1}{2}\eta^{\gamma\kappa}\eta^{\delta\iota}\epsilon_{\kappa\iota\mu\nu}\mathfrak{g}^{\mu}\wedge\mathfrak{g}^{\nu})\\
&=\boldsymbol{\delta}\mathfrak{g}^{\mu}\wedge(\eta^{\gamma\kappa}\eta^{\delta\iota}\epsilon_{\kappa\iota\mu\nu}\mathfrak{g}^{\nu}),
\end{aligned} \qquad (E.12)$$

from where we get

$$\frac{\partial\underset{g}{\star}(\mathfrak{g}^{\gamma}\wedge\mathfrak{g}^{\delta})}{\partial\mathfrak{g}^{\mathbf{m}}}=\eta^{\gamma\kappa}\eta^{\delta\iota}\epsilon_{\kappa\iota\mu\nu}\mathfrak{g}^{\nu}. \qquad (E.13)$$

On the other hand we have

$$\begin{aligned}
&\mathfrak{g}_{\mu}\lrcorner\underset{g}{\star}(\mathfrak{g}^{\gamma}\wedge\mathfrak{g}^{\delta})=\mathfrak{g}_{\mu}\underset{g}{\lrcorner}(\frac{1}{2}\eta^{\gamma\kappa}\eta^{\delta\iota}\epsilon_{\kappa\iota\rho\sigma}\mathfrak{g}^{\rho}\wedge\mathfrak{g}^{\sigma})\\
&=\eta^{\gamma\kappa}\eta^{\delta\iota}\epsilon_{\kappa\iota\mu\nu}\mathfrak{g}^{\nu}.
\end{aligned} \qquad (E.14)$$

Moreover, using the fourth formula in Eq.(2.102), we can write

$$\begin{aligned}
\frac{\partial[\underset{g}{\star}(\mathfrak{g}^{\gamma}\wedge\mathfrak{g}^{\delta})]}{\partial\mathfrak{g}^{\kappa}}&=\mathfrak{g}_{\kappa}\underset{g}{\lrcorner}\underset{g}{\star}(\mathfrak{g}^{\gamma}\wedge\mathfrak{g}^{\delta})\\
&=\underset{g}{\star}[\mathfrak{g}_{\kappa}\wedge(\mathfrak{g}^{\gamma}\wedge\mathfrak{g}^{\delta})]=\underset{g}{\star}(\mathfrak{g}^{\gamma}\wedge\mathfrak{g}^{\delta}\wedge\mathfrak{g}_{\kappa}).
\end{aligned} \qquad (E.15)$$

Finally,

$$\boldsymbol{\delta}\underset{g}{\star}(\mathfrak{g}^{\gamma}\wedge\mathfrak{g}^{\delta})=\boldsymbol{\delta}\mathfrak{g}^{\kappa}\wedge\underset{g}{\star}(\mathfrak{g}^{\gamma}\wedge\mathfrak{g}^{\delta}\wedge\mathfrak{g}_{\kappa}). \qquad (E.16)$$

Then, using Eq.(E.10) and Eq.(E.16) in Eq.(E.9) we get

$$\boldsymbol{\delta}\mathcal{L}_{eh}=\frac{1}{2}d[\boldsymbol{\delta}\omega_{\gamma\delta}\wedge\underset{g}{\star}(\mathfrak{g}^{\gamma}\wedge\mathfrak{g}^{\delta})]+\boldsymbol{\delta}\mathfrak{g}^{\kappa}\wedge[\frac{1}{2}\mathcal{R}_{\alpha\beta}\wedge\underset{g}{\star}(\mathfrak{g}^{\alpha}\wedge\mathfrak{g}^{\beta}\wedge\mathfrak{g}_{\kappa})]. \qquad (E.17)$$

Now, the curvature 2-form fields are given by $\mathcal{R}_{\alpha\beta} = \frac{1}{2} R_{\alpha\beta\gamma\kappa}\mathfrak{g}^{\gamma} \wedge \mathfrak{g}^{\kappa}$ where $R_{\alpha\beta\gamma\kappa}$ are the components of the Riemann tensor, and then

$$\begin{aligned}\frac{1}{2}\mathcal{R}_{\alpha\beta} \wedge \underset{g}{\star}(\mathfrak{g}^{\alpha} \wedge \mathfrak{g}^{\beta} \wedge \mathfrak{g}_{\mu}) &= -\frac{1}{2}\underset{g}{\star}[\mathcal{R}_{\alpha\beta} \underset{g}{\lrcorner} (\mathfrak{g}^{\alpha} \wedge \mathfrak{g}^{\beta} \wedge \mathfrak{g}_{\mu})] \\ &= -\frac{1}{4} R_{\alpha\beta\gamma\kappa} \underset{g}{\star} [(\mathfrak{g}^{\gamma} \wedge \mathfrak{g}^{\kappa}) \underset{g}{\lrcorner} (\mathfrak{g}^{\alpha} \wedge \mathfrak{g}^{\beta} \wedge \mathfrak{g}_{\mu})] \\ &= -\star(\mathcal{R}_{\mu} - \frac{1}{2}R\mathfrak{g}_{\mu}) = -\star\mathcal{G}_{\mu}, \end{aligned} \tag{E.18}$$

where the $\mathcal{R}_{\mu}$ are the Ricci 1-form fields and $\mathcal{G}_{\mu}$ are the Einstein 1-form fields. So, finally, we can write

$$\int \delta(\mathcal{L}_{eh} + \mathcal{L}_m) = \int \delta\mathfrak{g}^{\alpha} \wedge (-\star\mathcal{G}_{\alpha} + \frac{\partial\mathfrak{L}_m}{\partial\mathfrak{g}^{\alpha}}) = 0. \tag{E.19}$$

To have the equations of motion in the form of Eq.(6.25) (here with $m = 0$), i.e.,

$$d\underset{g}{\star}\mathcal{S}^{\gamma} + \underset{g}{\star}t^{\gamma} = -\underset{g}{\star}\mathcal{T}^{\gamma} \tag{E.20}$$

it remains to prove that

$$d\underset{g}{\star}\mathcal{S}^{\gamma} + \underset{g}{\star}t^{\gamma} = -\star\mathcal{G}^{\gamma}. \tag{E.21}$$

In order to do that we recall that $\mathcal{L}_{eh}$ may be written (recall Eq.(6.13) and Eq.(6.17)) as

$$\begin{aligned}\mathcal{L}_{eh} &= \mathcal{L}_g - d(\mathfrak{g}^{\alpha} \wedge \underset{g}{\star} d\mathfrak{g}_{\beta}) \\ &= -\frac{1}{2}d\mathfrak{g}^{\alpha} \wedge \underset{g}{\star} d\mathfrak{g}_{\alpha} + \frac{1}{2}\underset{g}{\delta}\mathfrak{g}^{\alpha} \wedge \underset{g}{\star}\underset{g}{\delta}\mathfrak{g}_{\alpha} + \frac{1}{4}d\mathfrak{g}^{\alpha} \wedge \mathfrak{g}_{\alpha} \wedge \underset{g}{\star}(d\mathfrak{g}^{\beta} \wedge \mathfrak{g}_{\beta}) - d(\mathfrak{g}^{\alpha} \wedge \underset{g}{\star} d\mathfrak{g}_{\beta}) \\ &= -\frac{1}{2}d\mathfrak{g}^{\alpha} \wedge \mathfrak{g}_{\beta} \wedge \underset{g}{\star}(d\mathfrak{g}^{\beta} \wedge \mathfrak{g}_{\alpha}) + \frac{1}{4}d\mathfrak{g}^{\alpha} \wedge \mathfrak{g}_{\alpha} \wedge \underset{g}{\star}(d\mathfrak{g}^{\beta} \wedge \mathfrak{g}_{\beta}) - d(\mathfrak{g}^{\alpha} \wedge \underset{g}{\star} d\mathfrak{g}_{\beta}) \\ &= \frac{1}{2}d\mathfrak{g}_{\alpha} \wedge \underset{g}{\star}\mathcal{S}^{\alpha} - d(\mathfrak{g}^{\alpha} \wedge \underset{g}{\star} d\mathfrak{g}_{\beta}) \end{aligned} \tag{E.22}$$

Then

$$\begin{aligned}\delta\mathcal{L}_g &= \delta\mathfrak{g}^{\alpha} \wedge \frac{\partial\mathcal{L}_g}{\partial\mathfrak{g}^{\alpha}} + \delta d\mathfrak{g}^{\alpha} \wedge \frac{\partial\mathcal{L}_g}{\partial d\mathfrak{g}^{\alpha}} \\ &\delta\mathfrak{g}^{\alpha} \wedge \left[\frac{\partial\mathcal{L}_g}{\partial\mathfrak{g}^{\alpha}} + d\left(\frac{\partial\mathcal{L}_g}{\partial d\mathfrak{g}^{\alpha}}\right)\right] + d\left(\delta\mathfrak{g}^{\alpha} \wedge \frac{\partial\mathcal{L}_g}{\partial d\mathfrak{g}^{\alpha}}\right) \\ &= \delta\mathfrak{g}^{\alpha} \wedge \left(\underset{g}{\star}t_{\alpha} + d\underset{g}{\star}\mathcal{S}_{\alpha}\right) + d(\delta\mathfrak{g}^{\alpha} \wedge \underset{g}{\star}\mathcal{S}_{\alpha}), \end{aligned} \tag{E.23}$$

with

$$\underset{g}{\star}t_{\alpha} := \frac{\partial\mathcal{L}_g}{\partial\mathfrak{g}^{\alpha}}, \quad \underset{g}{\star}\mathcal{S}_{\alpha} := \frac{\partial\mathcal{L}_g}{\partial d\mathfrak{g}^{\alpha}} \tag{E.24}$$

To calculate $\frac{\partial\mathfrak{L}_g}{\partial\mathfrak{g}^{\alpha}}$, we first recall from Eq.(2.23) that we can write

$$\mathfrak{g}_{\kappa}\underset{g}{\lrcorner}(\frac{1}{2}d\mathfrak{g}_{\alpha} \wedge \underset{g}{\star}\mathcal{S}^{\alpha}) = \frac{1}{2}(\mathfrak{g}_{\kappa}\underset{g}{\lrcorner} d\mathfrak{g}_{\alpha}) \wedge \underset{g}{\star}\mathcal{S}^{\alpha} + \frac{1}{2}d\mathfrak{g}_{\alpha} \wedge (\mathfrak{g}_{\kappa}\underset{g}{\lrcorner}\underset{g}{\star}\mathcal{S}^{\alpha}) \tag{E.25}$$

and thus since from Eq.(E.22) it is $\mathcal{L}_g = \frac{1}{2} d\mathfrak{g}_\alpha \wedge \underset{g}{\star} \mathcal{S}^\alpha$, we have

$$\mathfrak{g}_\kappa \lrcorner_g \mathcal{L}_g - (\mathfrak{g}_\kappa \lrcorner_g d\mathfrak{g}_\alpha) \wedge \underset{g}{\star} \mathcal{S}^\alpha = \frac{1}{2} \left[ d\mathfrak{g}_\alpha \wedge (\mathfrak{g}_\kappa \lrcorner_g \underset{g}{\star} \mathcal{S}^\alpha) - (\mathfrak{g}_\kappa \lrcorner_g d\mathfrak{g}_\alpha) \wedge \underset{g}{\star} \mathcal{S}^\alpha \right] \tag{E.26}$$

Next, we recall that

$$\begin{aligned} \boldsymbol{\delta} \mathcal{S}^\alpha &= \frac{1}{2} \mathcal{S}^\alpha_{\kappa\iota} \boldsymbol{\delta}(\mathfrak{g}^\kappa \wedge \mathfrak{g}^\iota) = \boldsymbol{\delta} \mathfrak{g}^\kappa \wedge (\mathfrak{g}_\kappa \lrcorner_g \mathcal{S}^\alpha), \\ \boldsymbol{\delta} \underset{g}{\star} \mathcal{S}^\alpha &= \boldsymbol{\delta} \mathfrak{g}^\kappa \wedge (\mathfrak{g}_\kappa \lrcorner_g \underset{g}{\star} \mathcal{S}^\alpha) \end{aligned} \tag{E.27}$$

and, so we have

$$\begin{aligned} \left[ \boldsymbol{\delta}, \underset{g}{\star} \right] \mathcal{S}^\alpha &= \boldsymbol{\delta} \underset{g}{\star} \mathcal{S}^\alpha - \underset{g}{\star} \boldsymbol{\delta} \mathcal{S}^\alpha \\ &= \boldsymbol{\delta} \mathfrak{g}^\kappa \wedge (\mathfrak{g}_\kappa \lrcorner_g \mathcal{S}^\alpha) - \boldsymbol{\delta} \mathfrak{g}^\kappa \wedge (\mathfrak{g}_\kappa \lrcorner_g \underset{g}{\star} \mathcal{S}^\alpha). \end{aligned} \tag{E.28}$$

Multiplying Eq.(E.28) on the left by $d\mathfrak{g}_\alpha \wedge$ and, moreover adding $\boldsymbol{\delta} d\mathfrak{g}^\kappa \wedge \underset{g}{\star} \mathcal{S}^\alpha$ to both sides, we get

$$\begin{aligned} \boldsymbol{\delta}(\frac{1}{2} \mathfrak{g}^\kappa \wedge \underset{g}{\star} \mathcal{S}^\alpha) &= \boldsymbol{\delta} d\mathfrak{g}^\kappa \wedge \frac{1}{2} \underset{g}{\star} \mathcal{S}^\alpha + \frac{1}{2} d\mathfrak{g}^\kappa \wedge \underset{g}{\star} \boldsymbol{\delta} \mathcal{S}^\alpha \\ &+ \boldsymbol{\delta} \mathfrak{g}^\kappa \wedge \frac{1}{2} \left( d\mathfrak{g}_\alpha \wedge (\mathfrak{g}_\kappa \lrcorner_g \underset{g}{\star} \mathcal{S}^\alpha) - (\mathfrak{g}_\kappa \lrcorner_g d\mathfrak{g}_\alpha) \wedge \underset{g}{\star} \mathcal{S}^\alpha \right) \end{aligned}$$

or

$$\begin{aligned} \boldsymbol{\delta} \mathcal{L}_g &= \boldsymbol{\delta} d\mathfrak{g}^\kappa \wedge \frac{1}{2} \underset{g}{\star} \mathcal{S}^\alpha + \frac{1}{2} d\mathfrak{g}^\kappa \wedge \underset{g}{\star} \boldsymbol{\delta} \mathcal{S}^\alpha \\ &+ \boldsymbol{\delta} \mathfrak{g}^\kappa \wedge \frac{1}{2} \left( d\mathfrak{g}_\alpha \wedge (\mathfrak{g}_\kappa \lrcorner_g \underset{g}{\star} \mathcal{S}^\alpha) - (\mathfrak{g}_\kappa \lrcorner_g d\mathfrak{g}_\alpha) \wedge \underset{g}{\star} \mathcal{S}^\alpha \right). \end{aligned} \tag{E.29}$$

Thus, comparing the coefficient of $\boldsymbol{\delta} \mathfrak{g}^\kappa$ in Eq.(E.29) with the one appearing on the first line in Eq.(E.23), and taking into account the identity given by Eq.(E.26), we get

$$\underset{g}{\star} t_\alpha := \frac{\partial \mathcal{L}_g}{\partial \mathfrak{g}^\alpha} = \mathfrak{g}_\alpha \lrcorner_g \mathfrak{L}_g - (\mathfrak{g}_\alpha \lrcorner_g d\mathfrak{g}^\kappa) \wedge \underset{g}{\star} \mathcal{S}_\kappa. \tag{E.30}$$

Also, take into account that

$$\underset{g}{\star} \mathcal{S}_\alpha := \frac{\partial \mathcal{L}_g}{\partial d\mathfrak{g}^\alpha} = -\mathfrak{g}_\kappa \wedge \underset{g}{\star} (d\mathfrak{g}^\kappa \wedge \mathfrak{g}_\alpha) + \tfrac{1}{2} \mathfrak{g}_\alpha \wedge \underset{g}{\star} (d\mathfrak{g}^\kappa \wedge \mathfrak{g}_\kappa), \tag{E.31}$$

or

$$\underset{g}{\star} \mathcal{S}^\kappa = -\underset{g}{\star} d\mathfrak{g}^\kappa - (\mathfrak{g}^\kappa \lrcorner_g \underset{g}{\star} \mathfrak{g}^\alpha) \wedge \underset{g}{\star} d \underset{g}{\star} \mathfrak{g}_\alpha + \frac{1}{2} \mathfrak{g}^\kappa \wedge \underset{g}{\star} (d\mathfrak{g}^\alpha \wedge \mathfrak{g}_\alpha). \tag{E.32}$$

Finally, comparing the coefficient of $\boldsymbol{\delta} \mathfrak{g}^\kappa$ in Eq.(E.29) with the one appearing on Eq.(E.17), we get the proof of Eq.(E.21).

**Remark E1** Before proceeding, recall that in our theory the $\star t^\alpha$ are 3-form fields defined directly as functions of the gravitational potentials $\mathfrak{g}^\alpha$ and the fields $d\mathfrak{g}^\alpha$ (Eqs.(6.26) and (6.27)), where the set $\{\mathfrak{g}^\alpha\}$ defines by chance a basis for the $\boldsymbol{g}$-orthonormal bundle. However, from Remark 6.4, we know that the cotetrad fields $\{\mathfrak{g}^\alpha\}$ defining the gravitational potentials are associated with the extensor field $\boldsymbol{h}$ modulus, an arbitrary local Lorentz rotation $\Lambda$ which is hidden in the definition of $\boldsymbol{g} = \boldsymbol{h}^\dagger\eta\boldsymbol{h} = \boldsymbol{h}^\dagger\Lambda^\dagger\eta\Lambda\boldsymbol{h}$. Thus the energy-momentum 3-forms $\underset{g}{\star}t^\gamma$ are gauge dependent. However, once a gauge is chosen the $\underset{g}{\star}t^\gamma$ are, of course, independent on the basis (coordinate or otherwise) where they are expressed, i.e., they are legitimate tensor fields. However, in *GRT,* where the $\underset{g}{\star}\mathcal{S}^\gamma : U \to \bigwedge^2 U$ are called *superpotentials*[2] and the $\underset{g}{\star}t^\gamma$ are called the *gravitational pseudo energy-momentum 3-forms* this is *not* the case. The reason is implicit in the name *pseudo* 3- *forms* for $\underset{g}{\star}t^\gamma$. Indeed, in *GRT,* those objects are directly associated with the connection 1-forms $\omega_{\alpha\beta}$ associated with a particular section of the $\mathbf{g}$-orthonormal bundle and thus, besides being gauge dependent, their components in any basis do *not* define a true tensor field.

Taking into account Remark E1, we provide alternative expressions for $\underset{g}{\star}t^\gamma$ and $\underset{g}{\star}\mathcal{S}^\gamma$ given by Eqs.(E.30) and (E.31) as

$$
\begin{aligned}
\underset{g}{\star}t^\gamma &= -\frac{1}{2}\omega_{\alpha\beta}\wedge[\omega^\gamma_\delta\wedge\underset{g}{\star}(\mathfrak{g}^\alpha\wedge\mathfrak{g}^\beta\wedge\mathfrak{g}^\delta) + \omega^\beta_\delta\wedge\underset{g}{\star}(\mathfrak{g}^\alpha\wedge\mathfrak{g}^\beta\wedge\mathfrak{g}^\gamma)],\\
\underset{g}{\star}\mathcal{S}^\gamma &= \frac{1}{2}\omega_{\alpha\beta}\wedge\underset{g}{\star}(\mathfrak{g}^\alpha\wedge\mathfrak{g}^\beta\wedge\mathfrak{g}^\gamma),
\end{aligned}
\tag{E.33}
$$

where we use Eq.(D.7) for packing part of the complicated formulas involving $\mathfrak{g}^\alpha$ and $d\mathfrak{g}^\alpha$, in the form of $\omega_{\alpha\beta}$. With Eqs.(E.33), we can provide a simple proof of Eq.(E.21). Indeed, let us compute $-2\underset{g}{\star}\mathcal{G}^\gamma$ using Eq.(E.18) and Cartan's second structure equation.

$$
\begin{aligned}
-2\underset{g}{\star}\mathcal{G}^\delta &= d\boldsymbol{\omega}_{\alpha\beta}\wedge\underset{g}{\star}(\mathfrak{g}^\alpha\wedge\mathfrak{g}^\beta\wedge\mathfrak{g}^\delta) + \boldsymbol{\omega}_{\alpha\gamma}\wedge\boldsymbol{\omega}^\gamma_\beta\wedge\underset{g}{\star}(\mathfrak{g}^\alpha\wedge\mathfrak{g}^\beta\wedge\mathfrak{g}^\delta)\\
&= d[\boldsymbol{\omega}_{\alpha\beta}\wedge\underset{g}{\star}(\mathfrak{g}^\alpha\wedge\mathfrak{g}^\beta\wedge\mathfrak{g}^\delta)] + \boldsymbol{\omega}_{\alpha\beta}\wedge d\underset{g}{\star}(\mathfrak{g}^\alpha\wedge\mathfrak{g}^\beta\wedge\mathfrak{g}^\delta)\\
&+ \boldsymbol{\omega}_{\alpha\gamma}\wedge\boldsymbol{\omega}^\gamma_\beta\wedge\underset{g}{\star}(\mathfrak{g}^\alpha\wedge\mathfrak{g}^\beta\wedge\mathfrak{g}^\delta)
\end{aligned}
$$

$$
\begin{aligned}
&= d[\boldsymbol{\omega}_{\alpha\beta}\wedge\underset{g}{\star}(\mathfrak{g}^\alpha\wedge\mathfrak{g}^\beta\wedge\mathfrak{g}^\delta)] - \boldsymbol{\omega}_{\alpha\beta}\wedge\boldsymbol{\omega}^\alpha_\rho\wedge\underset{g}{\star}(\mathfrak{g}^\rho\wedge\mathfrak{g}^\beta\wedge\mathfrak{g}^\delta)\\
&- \boldsymbol{\omega}_{\alpha\beta}\wedge\boldsymbol{\omega}^\beta_\rho\wedge\underset{g}{\star}(\mathfrak{g}^\alpha\wedge\mathfrak{g}^\rho\wedge\mathfrak{g}^\delta) - \boldsymbol{\omega}_{\alpha\beta}\wedge\boldsymbol{\omega}^\delta_\rho\wedge\underset{g}{\star}(\mathfrak{g}^\alpha\wedge\mathfrak{g}^\beta\wedge\mathfrak{g}^\rho)\\
&+ \boldsymbol{\omega}_{\alpha\gamma}\wedge\boldsymbol{\omega}^\gamma_\beta\wedge\underset{g}{\star}(\mathfrak{g}^\alpha\wedge\mathfrak{g}^\beta\wedge\mathfrak{g}^\delta)\\
&= d[\boldsymbol{\omega}_{\alpha\beta}\wedge\underset{g}{\star}(\mathfrak{g}^\alpha\wedge\mathfrak{g}^\beta\wedge\mathfrak{g}^\delta)] - \boldsymbol{\omega}_{\alpha\beta}\wedge[\boldsymbol{\omega}^\delta_\rho\wedge\underset{g}{\star}(\mathfrak{g}^\alpha\wedge\mathfrak{g}^\beta\wedge\mathfrak{g}^\rho)\\
&+ \boldsymbol{\omega}^\beta_\rho\wedge\underset{g}{\star}(\mathfrak{g}^\alpha\wedge\mathfrak{g}^\rho\wedge\mathfrak{g}^\delta)]\\
&= 2(d\underset{g}{\star}\mathcal{S}^\delta + \underset{g}{\star}t^\delta).
\end{aligned}
\tag{E.34}
$$

---

[2]These objects are called in [1], the Sparling forms [2]. However, they were already used much earlier by Thirring and Wallner in [3].

# References

[1] Szabados, L. B., Quasi-Local Energy-Momentum and Angular Momentum in GR: A Review Article, *Living Reviews in Relativity.* [http://www.livingreviews.org/lrr-2004-4]
[2] Sparling, G. A. J., *Twistors, Spinors and the Einstein Vacuum Equations* (unknown status), University of Pittsburg preprint (1982).
[3] Thirring, W., and Wallner, R., The Use of Exterior Forms in Einstein's Gravitational Theory, *Brazilian J. Phys.* **8**, 686-723 (1978).

# Appendix F

## Comment on the *LDG* Gauge Theory of Gravitation

In the footnote 1 of Chapter 1, we recalled that in *GRT*, a gravitational field is defined by an equivalence class of pentuples, where $(M, \mathbf{g}, D, \tau_g, \uparrow)$ and $(M', \mathbf{g}', D', \tau'_g, \uparrow')$ are said to be equivalent if there is a diffeomorphism $f : M \to M'$, such that $\mathbf{g}' = f^*\mathbf{g}$, $D' = f^*D$, $\tau'_g = f^*\tau_g, \uparrow' = f^* \uparrow$, (where $f^*$ here denotes the pullback mapping)[1]. Moreover, in *GRT*, when one is studying the coupling of the matter fields, represented, say by some sections of the exterior algebra bundle[2] $\psi_{1,\ldots,}\psi_n$, which satisfy certain coupled differential equations (*the equations of motion*[3]), two models $\{(M, \mathbf{g}, D, \tau_g, \uparrow), \psi_{1,\ldots,}\psi_n\}$ and $\{(M', \mathbf{g}', D', \tau'_g, \uparrow'), \psi'_{1,\ldots,}\psi'_n\}$ are said to be dynamically equivalent iff there is a diffeomorphism $f : M \to M'$, such that the structures $(M, \mathbf{g}, D, \tau_g, \uparrow)$ and $(M', \mathbf{g}', D', \tau'_g, \uparrow')$ are diffeormorphic and moreover $\psi'_i = f^*\psi_i$, $i = 1, \ldots, n$.

This kind of equivalence is not a peculiarity of *GRT*. It is an obvious mathematical requirement that any theory formulated with tensor fields existing in an arbitrary manifold must satisfy[4]. This fact is sometimes confused with the fact that, for a particular theory, say $\mathfrak{T}$, it may happen that there are diffeomorphisms $f : M \to M$ such that for a given model $\{(M, \mathbf{g}, D, \tau_g, \uparrow), \psi_{1,\ldots,}\psi_n\}$ it is $\mathbf{g}' = f^*\mathbf{g}$, $D' = f^*D$, $\tau'_g = f^*\tau_g, \uparrow' = f^* \uparrow$, but $\psi'_i \neq f^*\psi_i$. When this happens, the set of

---

[1] Details may be found in [1,2,3].

[2] If the exterior algebra bundle $\bigwedge T^*M$ is viewed as embedded in the Clifford algebra bundle $\mathcal{C}\ell(M, \mathbf{g})$ ($\bigwedge T^*M \hookrightarrow \mathcal{C}\ell(M, \mathbf{g})$) then even spinor fields can be represented in that formalism as some appropriate equivalence classes of sums of even non homogeneous differential forms. Details may be found in [1].

[3] The equations of motion, for any particular problem, satisfy appropriate initial and boundary conditions.

[4] Some people now call this property background independence, although we do not think that to be a very appropriate name. Others call this property general covariance. Eventually it is most appropriate to simply say that the theory is invariant under diffeomorphisms. Moreover, we recall that there is an almost trivial mathematical theorem saying that the *action functional* for the matter fields represented by sections of the exterior algebra bundle plus the gravitational field is invariant under the action of a one-parameter group of diffeomorphisms for Lagrangian densities that vanish on the boundary of the region where the action functional is defined. Details, for example, in [1].

diffeomorphisms satisfying that property defines a group, called a symmetry group of $\mathfrak{T}$, and of course, knowing the possible symmetry groups of a given theory facilitates the finding of solutions for the equations of motion [5].

Having said that, we now analyze first the motivations for the 'gauge theory of the gravitational field' in Minkowski spacetime of references [4] and [5] formulated on a vector manifold $U$ [6]. Those authors, taking advantage of the obvious identification of the tangent spaces of $U$, write that for a diffeomorphism $f : U \to U$, $x \mapsto f(x)$ we have some particularly useful representations for the pullback mapping ($f^*$, $f^*_{x'} : T^*_{x'}U \to T^*_x U$ and pushforward mapping ($f_*$, $f_{*_x} : T_xU \to T_{x'}U$). Indeed, for representing the pushforward mapping [1] and [2], let's introduce the operator $\mathtt{f}$, such that

$$\begin{array}{rcll} \mathtt{f} & : & \sec \bigwedge^1 TU & \to & \sec \bigwedge^1 TU, \\ & & a & \mapsto & \mathtt{f}(a) = a \underset{\eta}{\cdot} \partial_x f \end{array} \tag{F.1}$$

where

$$\partial_x = \gamma^\mu \frac{\partial}{\partial \mathrm{x}^\mu}, \tag{F.2}$$

with $\{\mathrm{x}^\mu\}$ coordinates for $U$ in the Einstein-Lorentz-Poincaré gauge and $\{\gamma^\mu\}$ the dual basis of the basis $\{\vartheta_\mu\}$, as introduced[7] in Chapter 5. For representing the pullback mapping [1] and [2] introduce the operator $\mathtt{f}^\dagger$ such that[8]

$$\begin{gathered} \underline{\mathtt{f}}^\dagger : \sec \bigwedge T_x U \to \sec \bigwedge T_x U, \\ \psi_i(x) \mapsto \psi_i'(x) = (\underline{\mathtt{f}}^\dagger \psi_i)(x) = \psi_i(f(x)) \end{gathered} \tag{F.3}$$

Next, those authors argue that the description of physics in terms of $\psi_i$ or $\psi_i^{'}$ is not *covariant.* To cure this 'deficiency', they propose to introduce a gauge field $\boldsymbol{h} : \sec \bigwedge^1 TU \to \sec \bigwedge^1 TU$, such that,

$$\begin{aligned} \psi_i(x) &= \boldsymbol{h}^\dagger [\Psi_i(x)], \\ \psi_i'(x) &= \boldsymbol{h}^{\dagger\prime} [\Psi_i(x)], \end{aligned} \tag{F.4}$$

where $\Psi_i(x)$ is said to be the covariant description of $\psi_i(x)$. This implies the following transformation law for $\boldsymbol{h}$ under a diffeomorphism $f : U \to U$,

$$\boldsymbol{h}^{\dagger\prime} = \underline{\mathtt{f}}^\dagger \boldsymbol{h}^\dagger. \tag{F.5}$$

From this point, using heuristic arguments, those authors say that this permits the introduction of a new metric extensor field on $U$, that they write as

$$\boldsymbol{g} = \boldsymbol{h}^\dagger \boldsymbol{h}, \tag{F.6}$$

because, different from what we did in this book they choose as basic algebra for performing calculations $\mathcal{C}\ell(U,\eta)$, instead of $\mathcal{C}\ell(U,\cdot)$. Also, take notice that in [4,5] it is $\boldsymbol{h} = h^{-1}$.

---

[5] See some examples, e.g., in Chapter 5 of [1].

[6] The vector manifod $U$ is as defined in Chapter 5, but with the elements of $U$ being vectors instead of forms, and the elements of $\bigwedge TU$ being multivector fields instead of multiform fields.

[7] Take notice that the reciprocal basis of $\{\gamma^\mu\}$ is the basis $\{\gamma_\mu\}$ such that $\gamma_\mu \underset{\eta}{\cdot} \gamma^\nu = \delta^\nu_\mu$.

[8] Recall that $\underline{\mathtt{f}}^\dagger$ is the extended of $\mathtt{f}$.

The reasoning behind Eqs.(F.4) is that the field $\Psi_i \in \sec \bigwedge TU$ has been selected as a representative of the class of the diffeomorphically equivalent fields $\{\underline{\mathbf{f}}^\dagger \psi_i\}$, for $f \in \mathrm{Diff}U$, the diffeomorphism group of $U$. This is a very important point to keep in mind. Indeed, in [4] and [5] in a region $V \subset U$ the gravitational potentials (called *displacement gauge invariant frame* $\{g^\mu\}$) are introduced as follows. First, take arbitrary coordinates $\{x^\mu\}$, covering $V$, and introduce the coordinate vector fields

$$e_\mu := \frac{\partial x}{\partial x^\mu} = \partial_\mu x, \tag{F.7}$$

where the position vector $x := \mathrm{x}^\mu(x^\alpha)\gamma_\alpha$ and the reciprocal basis of $\{e_\mu\}$ is the basis $\{e^\mu\}$ such that[9]

$$e^\mu := \partial_x x^\mu. \tag{F.8}$$

Then,

$$g_\mu := \boldsymbol{h}^\dagger(e_\mu) = h_\mu^\alpha e_\alpha, \tag{F.9}$$

$$g^\mu = \boldsymbol{h}^{-1}(e^\mu) = (\boldsymbol{h}^{-1}\partial_x)x^\mu \tag{F.10}$$

and

$$\boldsymbol{h}^\dagger(e_\mu) \underset{\eta}{\cdot} \boldsymbol{h}^\dagger(e_\nu) = \boldsymbol{g}(e_\mu) \underset{\eta}{\cdot} e_\nu = g_{\mu\nu}, \tag{F.11}$$

$$\boldsymbol{h}^{-1}(e^\mu) \underset{\eta}{\cdot} \boldsymbol{h}^{-1}(e^\nu) = \boldsymbol{g}^{-1}(e^\mu) \underset{\eta}{\cdot} e^\nu = g^{\mu\nu}. \tag{F.12}$$

From its definition, the commutator $[e_\mu, e_\nu] := e_\mu \underset{\eta}{\cdot} \partial_x e_\nu - e_\nu \underset{\eta}{\cdot} \partial_x e_\mu = 0$, but, in general ,we have that

$$\begin{aligned}[g_\mu, g_\nu] = [\boldsymbol{h}^\dagger e_\mu, \boldsymbol{h}^\dagger e_\nu] &:= (\boldsymbol{h}^\dagger e_\mu \underset{\eta}{\cdot} \partial_x)\boldsymbol{h}^\dagger e_\nu - (\boldsymbol{h}^\dagger e_\nu \underset{\eta}{\cdot} \partial_x)\boldsymbol{h}^\dagger e_\mu \\ &= (\partial_\mu h_v^\alpha - \partial_\nu h_\mu^\alpha)e_\alpha = c_{\mu\nu}^\alpha e_\alpha \neq 0, \end{aligned} \tag{F.13}$$

i.e., $\{g_\mu\}$ is *not* a coordinate basis, because the structure coefficients $c_{\mu\nu}^\alpha$ of the basis $g_\mu$ are non null.

**Remark F1** Before proceeding, we recall that on page 477 of [4] it is stated that the main property that must be required from a nontrivial field strength $\boldsymbol{h}$ is that we should not have $\boldsymbol{h}^\dagger(a) = \mathtt{f}^\dagger(a)$ where $\mathtt{f}$ is obtained from a diffeomorphism $f : U \to U$. This implies that $\partial_x \wedge \boldsymbol{h}^\dagger(e_\mu) \neq 0$, i.e., $(\partial_\mu h_v^\alpha - \partial_\nu h_\mu^\alpha)e_\alpha \neq 0$.

The next step in [4] towards the formulation of their gravitational theory is the introduction of what those authors call a 'gauge covariant derivative' that 'converts $\partial_\mu$ into a covariant derivative'. If $M \in \sec \bigwedge TU$ they define

$$\boxed{\mathbf{D}_\mu M = \partial_\mu + \Omega_\mu \times M,} \tag{F.14}$$

with $\Omega : \sec\bigwedge^1 TU \to \sec \bigwedge^2 TU$, $\Omega_\mu = \Omega(e_\mu)$. Before going on we must observe that the correct notation for $\mathbf{D}_\mu$,, in order to avoid confusion[10], must be $\mathbf{D}_{e_\mu}$, and we shall use it in what follows,

[9] Recall that $\partial_x = \gamma^\mu \frac{\partial}{\partial \mathrm{x}^\mu} = e^\mu \frac{\partial}{\partial x^\mu}$, with $e^\mu = \gamma^\alpha \frac{\partial x^\mu}{\partial \mathrm{x}^\alpha}$.

[10] In [5] Eq.(F.15) is written $D_\mu M = \partial_\mu + \omega(g_\mu) \times M$. We must immediately conclude that $\omega(g_\mu) = \omega(\boldsymbol{h}^\dagger(e_\mu)) = \Omega(e_\mu)$, i.e., $\omega = \boldsymbol{h}^{\clubsuit}\Omega$. Moreover from the notation used in [5] we should not infer that $D_\mu$ is representing $D_{g_\mu}$, because if that was the case then the defining equation for $D_{g_\mu}$ should read $D_{g_\mu} M = h_\mu^\alpha \partial_\mu + \Omega(g_\mu) \times M$.

thus to start, we write

$$\mathbf{D}_{e_\mu} M = \partial_\mu + \Omega_\mu \times M. \tag{F.15}$$

We also introduce the basis $\{\theta^\mu\}$ of $T^*U$ dual of the basis $\{e_\mu\}$ of $TU$ and also the basis $\{\zeta^\mu\}$ of $T^*U$ dual of the basis $\{g_\mu\}$ of $TU$ and write as in [4,5]:

$$\mathbf{D}_{e_\mu} g_\nu := L^\alpha_{\mu\nu} g_\alpha, \quad \mathbf{D}_{e_\mu} g^\alpha = -L^\alpha_{\mu\nu} g^\nu, \quad \mathbf{D}_{e_\mu} \zeta^\alpha = -L^\alpha_{\mu\nu} \zeta^\nu \tag{F.16}$$

We also write

$$\mathbf{D}_{e_\mu} e_\nu := \mathbf{L}^\alpha_{\mu\nu} g_\alpha, \quad \mathbf{D}_{e_\mu} e^\alpha = -\mathbf{L}^\alpha_{\mu\nu} e^\nu, \quad \mathbf{D}_{e_\mu} \theta^\alpha = -\mathbf{L}^\alpha_{\mu\nu} \theta^\nu. \tag{F.17}$$

Note now that, whereas the indices $\mu$ and $\nu$ in $\mathbf{L}^\alpha_{\mu\nu}$ refer to the same basis[11], namely $\{e_\mu\}$ the indices $\mu$ and $\nu$ in $L^\alpha_{\mu\nu}$ are hybrid[12], i.e., the first ($\mu$) corresponds to the basis $\{e_\mu\}$ whereas the second ($\nu$) corresponds to the basis $\{g_\nu\}$. This observation is very important, for indeed, if we write explicitly $\boldsymbol{\eta}$ and $\mathbf{g}$, the metric tensors which correspond to the metric extensors $\eta^{-1}$ and $\boldsymbol{g}^{-1}$, we have immediately from $g_\mu \underset{\eta}{\cdot} g_\nu = g_{\mu\nu}$ and $e_\mu \underset{g}{\cdot} e_\nu = g_{\mu\nu}$ that

$$\boldsymbol{\eta} = g_{\mu\nu} \zeta^\mu \otimes \zeta^\nu, \qquad \mathbf{g} = g_{\mu\nu} \theta^\mu \otimes \theta^\nu, \tag{F.18}$$

with $\{\zeta^\mu\}$ the dual basis of $\{g_\mu\}$ and $\{\theta^\mu\}$ the dual basis of $\{e_\mu\}$.

We then get

$$\begin{aligned} \mathbf{D}_{e_\mu} \boldsymbol{\eta} &= \mathbf{D}_{e_\mu} (g_{\alpha\beta} \zeta^\alpha \otimes \zeta^\beta) \\ &= (\partial_\mu g_{\alpha\beta} - g_{\rho\beta} L^\rho_{\mu\alpha} - g_{\alpha\rho} L^\rho_{\mu\beta}) \zeta^\alpha \otimes \zeta^\beta, \end{aligned} \tag{F.19}$$

i.e., the connection $\mathbf{D}$ will be metric compatible with $\boldsymbol{\eta}$ iff

$$\partial_\mu g_{\alpha\beta} - g_{\rho\beta} L^\rho_{\mu\alpha} - g_{\alpha\rho} L^\rho_{\mu\beta} = 0. \tag{F.20}$$

Now, in [4,5] Eq.(F.20) is trivially true[13], but it is interpreted by those authors as meaning that $\mathbf{D}$ is compatible with $\mathbf{g}$. The error in those papers arises due to the confusion with the hybrid indices (holonomic and non holonomic) in $L^\rho_{\mu\alpha}$, and we already briefly mentioned in [6]. And, indeed, a trivial calculation shows immediately that

$$\mathbf{D}_{e_\mu} \mathbf{g} \neq 0, \tag{F.21}$$

i.e., the connection $D$ has a non null *nonmetricity* tensor relative to $\mathbf{g}$. Indeed, we know from the results of Section 4.9.1 that the connection which is metrically compatible with $\boldsymbol{g}$ is $\mathbf{D}'$, the $\boldsymbol{h}$-deformation of $\mathbf{D}$ such that if $a, b \in \sec \bigwedge^1 TM$,

$$\mathbf{D}'_a b = \boldsymbol{h}^{-1}(\mathbf{D}_a \boldsymbol{h}(b)). \tag{F.22}$$

Another serious equivocation in [4,5] is the following. Those authors introduce a Dirac like operator, here written as $\boldsymbol{D} = g_\alpha \mathbf{D}_{e_\alpha}$. Then, it is stated (see, e.g., Eq.(131) in [5]) that

$$\boldsymbol{D} \wedge g^\alpha = \frac{1}{2}(L^\alpha_{\mu\nu} - L^\alpha_{\nu\mu}) g^\nu \wedge g^\mu \tag{F.23}$$

---

[11] Those indices are holonomic, since $\{e_\mu\}$ is a coordinate basis.

[12] The index $\mu$ is holonomic and the index $\nu$ is non holonomic. This has already been observed in [6].

[13] Indeed, we have for one side that $D_{e_\mu}(g_\alpha \underset{\eta}{\cdot} g_\beta) = \partial_\mu g_{\alpha\beta}$ and on the other side $D_{e_\mu}(g_\alpha \underset{\eta}{\cdot} g_\beta) = D_{e_\mu} g_\alpha \underset{\eta}{\cdot} g_\beta + g_\alpha \underset{\eta}{\cdot} D_{e_\mu} g_\beta = g_{\rho\beta} L^\rho_{\mu\alpha} + g_{\alpha\rho} L^\rho_{\mu\beta}$.

*is* the torsion tensor (more precisely the torsion 2-forms $\Theta^\alpha$). However, since the structure coefficients of the basis $\{g_\mu\}$ are non null, the correct expression for the torsion 2-form fields $\Theta^\alpha$ are, as well known (see. e.g., [1] )

$$\Theta^\alpha = \frac{1}{2}(L^\alpha_{\mu\nu} - L^\alpha_{\nu\mu} - c^\alpha_{\mu\nu})g^\nu \wedge g^\mu. \tag{F.24}$$

It follows that even assuming $L^\alpha_{\mu\nu} = L^\alpha_{\nu\mu}$ the connection $\mathbf{D}$ has torsion, contrary to what is claimed in [5]. Of course, a simple calculation shows that the Riemann curvature tensor of $\mathbf{D}$ is also non null.

The above comments clearly show that the gravitational theory presented in [4,5] introduces a connection $\mathbf{D}$ that is never compatible with $\mathbf{g}$ and moreover does not reduce to *GRT* even when it happens that $\Theta^\alpha = 0$, and as such that theory seems to be *unfortunately* inconsistent concerning its claims[14].

# References

[1] Rodrigues, W. A. Jr., and Capelas de Oliveira, E., *The Many Faces of Maxwell, Dirac and Einstein Equations. A Clifford Bundle Approach.* Lecture Notes in Physics **722**, Springer, Heidelberg, 2007.

[2] Sachs, R. K., and Wu, H., *General Relativity for Mathematicians*, Springer-Verlag, New York 1977.

[3] Rodrigues, W. A., Jr., and Rosa, M. A. F., The Meaning of Time in Relativity and Einstein's Later View of the Twin Paradox, *Found. Phys.* **19**, 705–724 (1989).

[4] Doran, C., and Lasenby, A., *Geometric Algebra for Physicists*, Cambridge Univ. Press, Cambridge 2003.

[5] Hestenes, D., Gauge Theory Gravity with Geometrical Calculus, *Found. Phys.* **35**, 903-969 (2005).

[6] Fernández, V. V, Moya, A. M., and Rodrigues, W. A. Jr., Covariant Derivatives on Minkowski Manifolds, in R. Ablamowicz and Fauser, B. (eds.), Clifford Algebras and their Applications in Mathematical Physics (Ixtapa-Zihuatanejo, Mexico 1999), vol.1, Algebra and Physics, *Progress in Physics* **18**, pp 373-398, Birkhäuser, Boston, Basel and Berlin, 2000.

[14] At http://www.mrao.cam.ac.uk/~clifford/publications/index.html there is a list of articles (published in very good journals) written by the authors of [4] and collaborators and based on their gravitational theory may be found.

# Appendix G
## Gravitational Field as a Nonmetricity Tensor Field

In this Appendix to further illustrate the point of view used in this book, we give a model for gravitational field of a *point* mass, where that field is represented by the nonmetricity tensor of a special connection. To start with, suppose that the world manifold $M$ is a 4-dimensional manifold diffeomorphic to $\mathbb{R}^4$. Let ,moreover, $(t, x, y, z) = (x^0, x^1, x^2, x^3)$ be global Cartesian coordinates for $M$.

Next, introduce on $M$ two metric fields:

$$\boldsymbol{\eta} = dt \otimes dt - dx^1 \otimes dx^1 - dx^2 \otimes dx^2 - dx^3 \otimes dx^3, \tag{G.1}$$

and

$$\begin{aligned}\mathbf{g} = &\left(1 - \frac{2m}{r}\right) dt \otimes dt \\ &- \frac{1}{r^2}\left(\frac{2m}{r-2m}\right)\sum_{i=1}^{3}(x^i)^2 dx^i \otimes dx^i \\ &- dx^1 \otimes dx^1 + \left(1 - \frac{2m}{r}\right)^{-1}(dx^2 \otimes dx^2 - dx^3 \otimes dx^3).\end{aligned} \tag{G.2}$$

In Eq. (G.2)

$$r = \sqrt{(x^1)^2 + (x^2)^2 + (x^3)^2}. \tag{G.3}$$

Now, introduce $(t, r, \vartheta, \varphi) = (x^0, x'^1, x'^2, x'^3)$ as the usual spherical coordinates for $M$. Recall that

$$x^1 = r \sin\vartheta \cos\varphi,\ x^2 = r \sin\vartheta \sin\varphi,\ x^3 = r\cos\vartheta \tag{G.4}$$

and the range of these coordinates in $\eta$ are $r > 0$, $0 < \vartheta < \pi$, $0 < \varphi < 2\pi$. For $\mathbf{g}$, the range of the $r$ variable must be $(0, 2m) \cup (2m, \infty)$.

As it can be easily verified, the metric $\mathbf{g}$ in spherical coordinates is:

$$\mathbf{g} = \left(1 - \frac{2m}{r}\right) dt \otimes dt - \left(1 - \frac{2m}{r}\right)^{-1} dr \otimes dr - r^2 (d\vartheta \otimes d\vartheta + \sin^2 \vartheta d\varphi \otimes d\varphi), \tag{G.5}$$

which we immediately recognize as the *Schwarzschild metric* of *GR*. Of course, $\boldsymbol{\eta}$ is a Minkowski metric on $M$.

As next step, we introduce two *distinct* connections, $\mathring{D}$ and $D$ on $M$. We assume that $\mathring{D}$ is the Levi-Civita connection of $\boldsymbol{\eta}$ in $M$ and $D$ is the Levi-Civita connection of $\mathbf{g}$ in $M$. Then, by definition (see, e.g., [1] for more details) the nonmetricities tensors of $\mathring{D}$ relative to $\boldsymbol{\eta}$ and of $D$ relative to $\mathbf{g}$ are null, i.e.,

$$\begin{aligned} \mathring{D}\boldsymbol{\eta} &= 0, \\ D\mathbf{g} &= 0. \end{aligned} \tag{G.6}$$

However, the nonmetricity tensor $\mathfrak{A}^\eta \in \sec T_3^0 M$ of $\mathring{D}$ relative to $\mathbf{g}$ is non null, i.e.,

$$\mathring{D}\mathbf{g} = \mathfrak{A}^\eta \neq 0, \tag{G.7}$$

and also the nonmetricity tensor $\mathbf{A}^g \in \sec T_3^0 M$ of $D$ relative to $\boldsymbol{\eta}$ is non null, i.e.,

$$D\eta = \mathfrak{A}^g \neq 0. \tag{G.8}$$

We, now, calculate the components of $\mathbf{A}^\eta$ in the coordinated bases $\{\boldsymbol{\partial}_\mu\}$ for $TM$ and $\{dx^\nu\}$ for $T^*M$ associated with the coordinates $(x^0, x^1, x^2, x^3)$ of $M$. Since $\mathring{D}$ is the Levi-Civita connection of the Minkowski metric $\boldsymbol{\eta}$, we have that

$$\begin{aligned} \mathring{D}_{\boldsymbol{\partial}_\mu} \boldsymbol{\partial}_\nu &= L^\rho_{\mu\nu} \boldsymbol{\partial}_\rho = 0, \\ \mathring{D}_{\boldsymbol{\partial}_\mu} dx^\alpha &= -L^\alpha_{\mu\nu} dx^\nu = 0. \end{aligned} \tag{G.9}$$

i.e., the connection coefficients $L^\rho_{\mu\nu}$ of $\mathring{D}$ in this basis are *null*. Then, $\mathfrak{A}^\eta = Q_{\mu\alpha\beta} dx^\alpha \otimes dx^\beta \otimes dx^\mu$ is given by

$$\begin{aligned} \mathfrak{A}^\eta = \mathring{D}\mathbf{g} &= \mathring{D}_{\boldsymbol{\partial}_\mu} \left( g_{\alpha\beta} dx^\alpha \otimes dx^\beta \right) \otimes dx^\mu \\ &= \left(\frac{\partial g_{\alpha\beta}}{\partial x^\mu}\right) dx^\alpha \otimes dx^\beta \otimes dx^\mu. \end{aligned} \tag{G.10}$$

To fix ideas, recall that for $Q_{100}$, it is,

$$\begin{aligned} Q_{100} &= \frac{\partial}{\partial x^1} \left( 1 - \frac{2m}{\sqrt{(x^1)^2 + (x^2)^2 + (x^3)^2}} \right) \\ &= -2m \frac{\partial}{\partial x^1} \left( \frac{1}{\sqrt{(x^1)^2 + (x^2)^2 + (x^3)^2}} \right) \\ &= \frac{2m x^1}{[(x^1)^2 + (x^2)^2 + (x^3)^2]^{\frac{3}{2}}} = \frac{2m x^1}{r^3}, \end{aligned} \tag{G.11}$$

which is *non null* for $x^1 \neq 0$. Note also that $Q_{010} = Q_{001} = 0$.

**Remark G1** Note that using Riemann normal coordinates $\{\xi^\mu\}$ covering $V \subset U \subset M$, naturally adapted to a reference frame $\mathbf{Z} \in \sec TV$[1] in free fall ($D_{\mathbf{Z}}\mathbf{Z} = 0$, $d\alpha \wedge \alpha = 0$, $\alpha = g(\mathbf{Z}, \ )$), it is possible, as it is well known, to put the connection coefficients of the Levi-Civita connection $D$ of $\mathbf{g}$ equal to zero in all points of the world line of a free fall observer (an *observer* is here modelled as an integral line $\sigma$ of a reference frame $\mathbf{Z}$, where $\mathbf{Z}$ is a time like vector field pointing to the future such that $\mathbf{Z}|_\sigma = \sigma_*$).

In the Riemann normal coordinates system covering $U \subset M$, not all the connection coefficients of the non metric connection $\mathring{D}$ (relative to $g$ ) are null. Also (of course) the nonmetricity tensor $\mathfrak{A}^\eta$ of $\mathring{D}$ (relative to $\mathbf{g}$) representing the true gravitational field is not null. An observer following $\sigma$ does not feel any force along its world line because the gravitational force represented by the nonmetricity field is compensated by an inertial force represented by the non null connection coefficients[2] $L''^{\rho}_{\mu\nu}$ of $\mathring{D}$ in the basis $\{\frac{\partial}{\partial \xi^\nu}\}$.

The situation is somewhat analogous to what happens in any non inertial reference frame which, of course, may be conveniently used in *any* Special Relativity problem (as e.g., in a rotating disc [2]), where the connection coefficients of the Levi-Civita connection of $\boldsymbol{\eta}$ are not all null.

Standard clocks and standard rulers (standard according to *GRT*) are predicted to have a behavior that seems to be (at least with the precision of the *GPS* system) exactly the observed one. However, take into account that those objects are *not* the standard clocks and standard rulers of the model proposed here[3]. We would say that the gravitational field *distorts* the period of the standard clocks and the length of the standard rulers of *GRT* relative to the standard clocks and standard rulers of the proposed model, which has non null nonmetricity tensor[4]. But, who are the devices that materialize those concepts in the proposed model? Well, they are *paper concepts*, like the notion of time in some Newtonian theories. They are defined and calculated in order to make *correct* predictions. However, given the status of present technology, we can easily imagine how to build devices to directly realize the standard rulers and clocks of the proposed model.

# References

[1] Rodrigues, W. A. Jr., and Capelas de Oliveira, E., *The Many Faces of Maxwell, Dirac and Einstein Equations. A Clifford Bundle Approach*, Lecture Notes in Physics **722**, Springer, Heidelberg, 2007.

[2] Rodrigues, W. A. Jr., and Sharif, M., Rotating Frames in RT: Sagnac's Effect in SRT and other Related Issues, *Found. Phys.* **31**, 1767-1784 (2001).

[3] Coll, B., A Universal Law of Gravitational Deformation for General Relativity, *Proc. of the Spanish Relativistic Meeting, EREs*, Salamanca, Spain,1998.

[1] For the mathematical definitions of reference frames, naturally adapted coordinates to a reference frame and observers, see. e.g., Chapter 5 of [1].

[2] $\nabla_{\frac{\partial}{\partial \xi^\mu}} \frac{\partial}{\partial \xi^\nu} = L''^{\rho}_{\mu\nu} \frac{\partial}{\partial \xi^\nu}$. The explicit form of the coefficients $L''^{\rho}_{\mu\nu}$ may be found in Chapter 5 of [1].

[3] Of course, our intention was only to illustrate a statement we made in the Introduction, namely that some gravitational fields may have many (equivalent) geometrical interpretations.

[4] On his respect see also [3].

# Acronyms and Abbreviations

| | |
|---|---|
| ADM | Arnowitt-Deser-Misner |
| GRT | General Relativity Theory |
| GSS | Geometrical Space Structure |
| MCGSS | Metrical Compatibale Geometrical Space Structure |
| MCPS | Metrical Compatible Parallelism Structure |
| DHSF | Dirac-Hestenes Spinor Field |
| MECM | Multiform and Extensor Calculus on Manifolds |
| iff | if and only if |
| i.e. | Id Est |
| e.g. | Exempli Gratia |

# List of Symbols

$M$ manifold
$\boldsymbol{h}$ distortion field
$\boldsymbol{h}^{\dagger}$ adjoint of $\boldsymbol{h}$
$\boldsymbol{h}^{\clubsuit}$ inverse of the adjoint of $\boldsymbol{h}$
$\eta$ Minkowski metric extensor
$\boldsymbol{\eta}$ Minkowski metric tensor
$\boldsymbol{g}$ Lorentzian metric extensor
$\mathbf{g}$ Lorentzian metric tensor
$\bigwedge T^*M$ bundle of non homogenous differential forms
$\bigwedge^k T^*M$ bundle of homogeneous differential $k$-forms
sec section
$D, \nabla$ connections
$\tau_g$ volume element
$\{\vartheta^{\mu}\}$ orthonormal cobasis of Minkowski spacetime
$\{\mathfrak{g}^{\mu}\}$ gravitational potentials
$\uparrow$ time orientation
$\mathbf{T}_p$ particles energy-momentum tensor
$\mathbf{T}_f$ fields energy-momentum tensor
$\mathfrak{A}$ nonmetricity tensor
$\mathfrak{R}$ Riemann curvature tensor
$\Theta$ torsion tensor
$\mathcal{A}^n$ $n$-dimensional affine space
$\{\mathrm{x}^{\mu}\}$ coordinates in Einstein-Lorentz-Poincaré gauge
$\pounds_{\xi}$ Lie derivative in the direction of the $\xi$ vector field
$\mathbf{V}$ $n$-dimensional vector space

$V = \mathbf{V}^*$ dual space of $\mathbf{V}$
$\mathbb{R}$ real field
$\mathbb{C}$ complex field
$\bigwedge^k V$ space of homogenous $k$-forms
$\bigwedge V$ space of non homogenous forms
$\wedge$ exterior product
$\langle\ \rangle_k$ $k$-part operator
$\widehat{\ }$ principal involution operator
$\widetilde{\ }$ reversion operator
$\cdot$ canonical euclidean scalar product
$\underset{\eta}{\cdot}$ Minkowski scalar product
$\underset{g}{\cdot}$ metric scalar product
$\lrcorner, \llcorner$ canonical contractions
$\underset{g}{\lrcorner}, \underset{g}{\llcorner}$ metric contractions
$extV$ space of general extensor over $V$
$(p,q)$-$extV$ space of $(p,q)$-extensors over $V$
$\underline{t}$ exterior power extension of a $(1,1)$-extensor
$t_+$ symmetric part of a $(1,1)$-extensor
$t_-$ antisymmetric part of a $(1,1)$-extensor
$tr[t]$ trace of a $(1,1)$-extensor
$\det[t]$ determinant of a $(1,1)$-extensor
$T$ generalization of a $(1,1)$-extensor $t$
bif biform mapping
$\times$ canonical commutator
$\underset{g}{\times}$ metric commutator
$\mathcal{C}\ell(V,\cdot)$ canonical Clifford algebra
$\mathcal{C}\ell(V,\boldsymbol{g})$ metric Clifford algebra
$\star$ canonical Hodge star operator
$\underset{g}{\star}$ metric Hodge star operator
$A\cdot\partial_X$ $A$-directional derivative of multiform functions
$\partial_X$ derivative mapping of multiform functions
$\underline{t}(\partial_X)$ $t$-distortion of $\partial_X$
$\mathcal{F}_{(X_1,\ldots,X_k)}[t]$ multiform functional
$A\cdot\partial_t$ $A$-directional derivative of a functional
$\partial_t$ derivative of a functional
$\partial_t *$ $*$-product derivative of a functional
$\delta_t^w$, $\delta$ variational operator
$\gamma_{(x)}, \gamma$ connection 2-extensor field
$\gamma_a$ connection $(1,1)$-extensor field
$\Gamma_a$ generalization of $\gamma_a$
$\omega(a)$ rotation gauge field
$\nabla_a$, $\nabla_a^-$ covariant derivatives of multivector and extensor fields
$\overset{\gamma}{\tau}$ torsion 2-extensor field associated to a given connection $\gamma$

$\overset{\gamma}{T}$ torsion $(2,1)$-extensor field associated to a given connection $\gamma$
$\overset{\gamma}{\rho}$ curvature operator associated to a given connection $\gamma$
$\overset{\gamma}{\mathbf{R}}_1$ Riemann (curvature) 4-extensor field associated to given a connection $\gamma$
$\overset{\gamma}{\mathbf{R}}_2$ Ricci 2-extensor field
$\overset{\gamma}{\boldsymbol{R}}_1$ Ricci $(1,1)$-extensor field
$[a,b,c]$ first Christofell operator
$[a,b]$ Lie bracket of 1-form fields
$\left\{ \begin{matrix} c \\ a,b \end{matrix} \right\}$ second Christofell operator
$\lambda$ Levi-Civita connection on $U_o$
$\overset{\gamma g}{\omega}(a)$ $\gamma \boldsymbol{g}$-gauge rotation field
$\overset{\gamma g}{\mathbf{R}}_3$ Riemann (curvature) 4-extensor field associated to structure $(U_o, \boldsymbol{g}, \gamma)$
$\overset{\gamma g}{\mathbf{R}}_2$ Ricci 2-extensor field associated to structure $(U_o, \boldsymbol{g}, \gamma)$
$\overset{\gamma g}{R}$ Ricci scalar field
$\overset{\gamma g}{\boldsymbol{R}}_2$ Riemann $(2,2)$-extensor field associated to structure $(U_o, \boldsymbol{g}, \gamma)$
$\overset{\gamma g}{\boldsymbol{R}}_1$ Ricci $(1,1)$-extensor field associated to structure $(U_o, \boldsymbol{g}, \gamma)$
$\overset{\gamma g}{\boldsymbol{G}}$ Einstein $(1,1)$-extensor field associated to structure $(U_o, \boldsymbol{g}, \gamma)$
$D_a$, $D_a^-$ Levi-Civita covariant derivatives associated to $\lambda$
$\boldsymbol{R}_2$ Riemann $(2,2)$-extensor field associated to structure $(U_o, \boldsymbol{g}, \lambda)$
$\boldsymbol{R}_1$ Ricci $(1,1)$-extensor field associated to structure $(U_o, \boldsymbol{g}, \lambda)$
$\underset{g}{\boldsymbol{D}}$ $\boldsymbol{g}$-gradient associated to $D_a^-$
$\underset{g}{\boldsymbol{D}}\lrcorner$ $\boldsymbol{g}$-divergent associated to $D_a^-$
$\boldsymbol{D}\wedge$ rotational associated to $D_a^-$
$\chi$ connection compatible with Minkowski metric $\eta$
$\overset{\chi\eta}{\mathcal{D}}_a$, $\overset{\chi\eta-}{\mathcal{D}}_a$ $\eta$-metric compatible covariant derivatives
$\overset{\chi\eta}{\otimes}(a)$ $\chi\eta$-gauge rotation field associated to structure $(U_0, \eta, \chi)$
$\overset{\chi\eta}{\nabla}_a$, $\overset{\chi\eta-}{\nabla}_a$ $\boldsymbol{h}$-distortions of the $\overset{\chi\eta}{\mathcal{D}}_a$, $\overset{\chi\eta-}{\mathcal{D}}_a$ covariant derivatives
$\overset{\chi\boldsymbol{h}}{\mathcal{R}}_2$ Riemann ($(2,2)$-extensor) gauge field for $(U_0, \boldsymbol{g}, \gamma)$
$\overset{\chi\boldsymbol{h}}{\mathcal{R}}_1$ Ricci ($(1,1)$-extensor) gauge field for $(U_0, \boldsymbol{g}, \gamma)$
$\overset{\chi\boldsymbol{h}}{\mathcal{R}}$ Ricci (scalar) gauge field for $(U_0, \boldsymbol{g}, \gamma)$
$\overset{\chi\boldsymbol{h}}{\mathcal{G}}_1$ Einstein ($(1,1)$-extensor) gauge field for $(U_0, \boldsymbol{g}, \gamma)$
$\Omega(a)$ $\eta$-rotation operator for the Minkowski-Cartan structure $(U_0, \eta, \mu)$
$\mathfrak{L}_{eh}$ Einstein-Hilbert Lagrangian
$\mathcal{L}_{eh}$ Einstein-Hilbert Lagrangian density
$\mathcal{L}_g$ Lagrangian density for the gravitational field in terms of potentials
$\mathcal{L}_h$ $\boldsymbol{h}$-deformation of $\mathcal{L}_g$
$\mathcal{S}^\alpha$ superpotential 2-form fields
$\mathcal{T}^\alpha$ energy-momentum 1-form fields

$t^a$ gravitational energy-momnetum 1-form fields
$d$ differential operator
$\delta$ Hodge coderivative operator
$\underset{g}{\star}\mathbf{T}^{\alpha}$ energy-momentum 3-forms fields
$\underset{g}{\star}\mathbf{L}^{\alpha\beta}$ orbital angular-momentum 3-forms fields
$\underset{g}{\star}\mathbf{S}^{\alpha\beta}$ spin angular-momentum 3-forms fields
$\underset{g}{\star}\mathbf{J}^{\alpha\beta}$ total angular-momentum 3-forms fields
$\overset{\circ}{\Diamond}$ Hodge Laplacian
$\Box = \boldsymbol{\partial}\cdot\boldsymbol{\partial}$ covariant D'Alembertian operator
$\boldsymbol{\partial} \wedge \overset{g}{\boldsymbol{\partial}}$ Ricci operator

# Index

MIX
Papier aus verantwortungsvollen Quellen
Paper from responsible sources
FSC® C105338

If you have any concerns about our products,
you can contact us on
**ProductSafety@springernature.com**

In case Publisher is established outside the EU,
the EU authorized representative is:
**Springer Nature Customer Service Center GmbH**
**Europaplatz 3, 69115 Heidelberg, Germany**

Printed by Libri Plureos GmbH
in Hamburg, Germany